RUST

How to Keep It From Destroying Your Car

Steven B. Joseph

Ward Hill Press, Staten Island, NY

Illustrations by Diana Yates.

Cover design by Suzanne Lobel.

Published by Ward Hill Press

40 Willis Avenue

Staten Island, New York 10301

Library of Congress Catalog Card Number: 89-84792

ISBN 0-9623380-4-4

ACKNOWLEDGMENTS

This book would not have been possible without Jim Hein, who served as technical consultant. I am also indebted to Cliff Clark for all his research, and to Dennis Rodriguez, who spent many hours explaining computers and related matters. I also appreciate Derek Drew, who gave me excellent editorial advice.

Warning — Disclaimer

The author and publisher have tried very hard to make this book as complete and accurate as possible. However, there may be mistakes — both in typography and content.

Because this book is intended as an overview of rust prevention and treatment, it should be used as a general guide to the processes it describes, not as an infallible authority on body work, welding and rust prevention.

Neither the publisher nor the author takes responsibility for the safety or utility of the ideas presented here. The reader should use common sense when applying any techniques suggested in these pages. Furthermore, the reader should seek the assistance of professional body mechanics and welders where questions arise.

Finally, whenever any chemicals and machinery are used, the product's label or instructions should *always* be consulted; any warnings they contain should be heeded.

Table of Contents

Chapter 1
The Basics

*Rust: an electrochemical reaction that occurs
when iron reverts to its natural state as an oxide.*

In the world of cars, the deadliest disease is rust. Infected bodies are visible at a glance — blistered or freckled with reddish spores, riddled with jagged holes, even missing fenders, panels or bumpers.

Cars are especially vulnerable to rust because they are made of thin sheets of steel. (Higher-grade steel costs the carmakers too much money to be cost-effective.) Even today, with the carmakers applying new technology, the problem of auto body rust persists — primarily because the level of pollutants and corrosive elements in the environment skyrockets every year.

For example, did you know that the United States dumps more salts on its roads that does any other country in the world — more than 20 billion pounds a year? And that acid rain is spreading to encompass isolated, rural areas — not just large metropolitan centers? And did you know that American carmakers warrantee new cars only against inside-out corrosion — and then only when the car owner meets a list of requirements for care?

In addition, did you know that many new cars involved in traffic accidents are "clipped" in half and welded to other front or rear ends — a dangerous process that can lead to deadly corrosion? That a large percentage of cars on new-car lots have already been damaged and repainted, leaving them vulnerable to rust? That a car's working parts are just as susceptible to corrosion as its body panels? That after-market rustproofing is often a big waste of money? That some new chemicals on the market can add years to your car's life?

In truth, auto body rust is a pervasive and costly by-product of car ownership today, whether your car is new, used or "classic." There are many causes besides the ones listed above. And there are just as many approaches to its prevention and cure.

If you own a car today — any car — you owe it to yourself (and your wallet) to keep it in sound condition as long as possible. For one thing, you'll get more money when you decide to sell it or trade it in. But more important, a sound vehicle means safety and reliability.

So plot a strategy for prevention (Chapter 3). Acquire the skills necessary to make basic repairs (Chapter 4). Learn what your car's manufacturer owes you in the way of repair and protection (Chapter 10). Take the steps necessary to find a reliable body mechanic (Chapter 7). Protect your collectable car (Chapter 5).

Know how to tell the difference between effective after-market rustproofing and simple fraud (Chapter 9). Be thorough when you shop for a new or used car (Chapter 8). Know what the federal government will do for you (Chapter 11) and the basic federal and state laws governing manufacturing and rustproofing (Chapters 8 and 9). And, finally, keep up with the latest products and coatings that can help you fight or prevent rust (Chapter 6 and Appendix 1).

But first, know the basic causes of rust, and the most vulnerable areas on cars today (and yesterday).

Road salts. Road crews dump these chemicals in larger amounts each year in the United States, and increasingly in Europe. Calcium chlorides — used to control dust on dirt roads — fall into the same category. When they are wet, salts adhere to the underbody and exterior of your car, where they can remain for months. And it takes only a drop of moisture to reactivate their corrosive nature.

In the United States, the "Salt Belt," where the largest amounts of road salts are dumped each winter, includes 24 states and the District of Columbia: Connecticut, Delaware, Illinois, Indiana, Iowa, Kansas, Kentucky, Maine, Maryland, Massachusetts, Michigan, Minnesota, Missouri, Nebraska, New Hampshire, New Jersey, New York, Ohio, Pennsylvania, Rhode Island, Vermont, Virginia, West Virginia and Wisconsin.

Acid rain. Although it is most common in industrial regions, acid rain has spread to encompass rural areas. It occurs when airborne pollutants — primarily sulfuric and nitric acids — travel through rain, snow or fog. Alone acid rain is among the most corrosive elements. Combined with road salts, however, it becomes the single greatest factor contributing to automobile rust. For this reason, the most corrosive region in the United States is the Great Lakes region, where acid rain and road salt use abound.

Breaks in a car's surface include nicks or scratches, cracked undercoating, and stress cracks (which develop after a car is driven a few hundred miles). They are failures in the protective coating. Most outside-in perforation occurs because of such surface imperfections.

Mud, snow and slush are also bad news. They can contain chemicals (such as road salts), penetrate crevices in a car's underbody, and adhere to the surface for a long time.

Pollution. In areas of high industrial density, sulfuric acids in the air can damage a good paint finish if it goes unprotected.

High humidity and warm temperatures, especially along the coast, where salt air compounds the problem, can accelerate the corrosion rate.

Silt, sand and airborne debris also collect on a paint finish and in the crevices and cracks that develop. Untreated, they can result in paint blistering.

Dents can foster rust as well. Like nicks and scratches, they damage a car's protective coatings. Dents can also hold salt or silt deposits and moisture. Often, rust appears shortly after a fender becomes dented, and spreads rapidly.

Imprudent metal combinations encourage galvanic corrosion. One metal will disintegrate at a faster rate. Often, because of hurried repairs, a nut or bolt will remain intact at the expense of an entire fender.

Disuse or neglect. If a car is left sitting for weeks or months under improper conditions, rust may invade. For example, a car left in bright sunlight will develop faulty weatherstripping and end up with damaging water leaks. Condensation may form in a car with an empty or near-empty gas tank, causing rust and scale in the tank. Likewise, nicks or scratches left bare can become a network of corrosion.

VULNERABLE AREAS

Some parts of a car are more prone to rust than others. Areas nearer the ground, such as rocker panels and the underbody itself, are the first to sustain damage. Pebbles, dirt, mud and slush get kicked up by the tires during travel. Similarly, horizontal parts of the car are more likely to trap and hold mud and salts. Below is a list of other vulnerable areas.

Hard-to-reach or inaccessible areas, like inner wheel wells, box sections, closed rocker panels, pockets and crevices, the windshield and rear window borders, headlight housings, sharp angles between two pieces of metal, and the roll at the bottom of a curved fender also develop rust more quickly. It is harder to keep such areas free of salt and mud or to coat them with protective chemicals. Moreover, a car owner is less likely to notice damage in these areas until that damage has spread to more visible parts of the car.

Underbody areas, where soundproofing or undercoating materials have cracked or begun to flake, are especially vulnerable. Moisture and dirt or salt can gather between such materials and the car.

Repaired areas. A bad repair can cause more problems than it solves — if the repair-person damages adjacent panels, uses poor-quality replacement parts, welds at excessive temperatures, combines the wrong metals, or covers rusted areas with body filler without removing the rust.

Welded joints and open seams rust more quickly, especially when they involve two different metals. Joints between metal and nonmetal parts are also prone to rust because they usually contain crevices or gaps where moisture can collect.

Doors. Doors are especially vulnerable at mounting areas and along the interior bottom (because of drainage problems).

Floors are usually one of the first areas to rust completely through because corrosive elements threaten from below and above. They threaten from below because of cracked, moisture-infiltrated undercoating, or corroding box sections. They threaten from above because of interior body leaks, snow (brought in on footwear), etc.

Braking systems are prone to rust because of the temperature extremes under which they operate. In most cars, the braking system is an appendage to the basic design, not an integral part of the car. As a result, it is exposed to a wide range of corrosive elements. In addition, conventional brake fluids absorb moisture.

The cooling system rusts easily, primarily because of antifreeze left for several seasons, which loses its rust-inhibiting qualities. Rust deposits develop, which restrict the performance of the heater and radiator. In addition, aluminum cylinder heads may begin to erode into particles, which can build up into restrictive deposits.

The fuel and exhaust systems, where condensation can develop, are prone to rust and scale.

The ignition system corrodes easily, including spark plug wires, distributor cap terminals and the rotor.

The underside of front and rear bumpers form another trap for road salts, mud and slush.

Wheel studs are susceptible to rust, because of mud and slush kicked up by the tires.

Battery compartments, including tie-down brackets and bolts, also corrode quickly, usually because of leaky battery acid and temperature extremes.

Electrical connections. When these corrode, they can cause headlights to dim or malfunction. Sometimes they are even misdiagnosed as computer or fuel-injection problems.

See Chapter 2 for the latest in factory anti-rust technology. And plan a well-rounded approach to prevention and repair in the pages that follow.

Chapter 2
The Carmakers

Twenty or 30 years ago, cars were designed with short life spans. Gas was cheap. Rust wasn't that much of a problem. If it got to be, you just went out and bought a new machine. But nowadays, with new cars averaging $14,000 and the U.S. dumping record amounts of road salts every winter, and with acid rain and the like, the public has begun demanding a hardier product. As a result, the carmakers have been forced to deal with a subject they have managed for so long to avoid.

Because the U.S. dumps more road salts than any other country in the world, rusting cars are a bigger problem here than anywhere else. Consequently, American carmakers are a little further along in anti-rust technology, except for a few foreign pioneers like Mercedes-Benz, Volkswagen, Saab and Volvo.

THE NEW TECHNOLOGY

Galvanization. As I explained in Chapter 1, rust develops when iron and oxygen react in a moist environment. So one way to prevent rust is to erect a barrier between the iron and the oxygen (the primer and paint coats form such a barrier). Much of the new technology centers on this idea. Take galvanized steel, for example.

In the galvanization process, steel is coated with zinc or zinc-iron or a similar combination. Because zinc is more susceptible to corrosion than steel is, the zinc coating gives itself up when the car is exposed to corrosive elements like salt or pollutants. And the steel is shielded from oxygen and moisture.

Most of the galvanized steel used so far by the carmakers has been galvanized using the "hot-dip" process: the steel is literally dipped in vats of the zinc solution. One problem with this approach is that the metal surface often gets coated unevenly, resulting in imperfections that do not occur with cold-rolled steel. In the *electrogalvanization* process, however, the steel is given an electrical charge and then coated with a zinc solution of the opposite charge. The coating is more uniform, and the surface is easier to paint.

With electrogalvanization, the steel can be coated on one or two sides, depending on the application. It can even become "differentiated," meaning it is coated on one and a half sides, or in some other irregular manner. This is important because — however the steel is galvanized — it just doesn't hold paint as well as cold-rolled steel. Because of this, the carmakers use one-sided galvanized steel for areas of the car that are in the direct line of sight, such as hoods and fenders. The galvanization is on the inside, out of sight, and the regular steel is on the outside, where the paint is applied.

But the use of any galvanized steel complicates the manufacturing process because galvanized steel is thicker and more difficult to shape, weld and bond. Consequently, as they begin to use larger amounts of galvanized steel, the carmakers must adapt their designs and assembly procedures to the extra thickness.

When the 1987 model year rolled around, the steelmakers had access to much-improved galvanization facilities (primarily electrogalvanization facilities). And the amount of galvanized steel used on each car will continue to increase dramatically.

Drainage. The carmakers have also redesigned areas known to trap and hold moisture. These include seams, laps and joints. These areas are now being shaped so that water drains away from them properly. In fact, proper drainage throughout the car is one of the carmakers' primary goals.

Adhesives and composite materials. The manufacturers are also stepping up the use of adhesives and composite materials. With adhesives, the carmakers can begin to phase out welding, which becomes more difficult with galvanized steel. When two pieces of metal are bonded with adhesives, they form a continuous joint (instead of a series of small joints, as with spot welding). The connection is stronger, and there are fewer crevices and gaps to trap moisture and foster corrosion. In addition, with new technology for the rapid curing of adhesives, the carmakers may soon use them on a much larger scale. (Most adhesive bonding so far has involved only the nonstructural components of an automobile.)

Much automobile trim is now attached with adhesives. This eliminates the need to drill holes, which damage the metal's protective coatings and cause leakage problems, too. Chrysler bonds all of its doors (which consist of two pieces of metal). General Motors confines its use of adhesives primarily to a line of midsize cars.

Composites, such as glass-reinforced polyester, are seeing greater use. So are thermoplastics, like Azdel, which has a polypropylene base. These materials are now used for nonstructural automotive parts, such as bumper beams and battery supports and trays. They are also often used for trim and accessories.

In the future, composites and thermoplastics may be used for dashboards, doors, exterior body panels, fenders, floor pans and rear floors. In fact, one company, Carron & Co. of Inkster, Michigan, has already developed an all-plastic "concept car." This car is part of the company's research and development of *structural reactive injection molding* (SRIM), a technology for the production of composite materials for structural areas.

Composites and thermoplastics, once fully developed, will have definite advantages over steel. They are lighter and resist corrosion. They are also easy to bond and fasten. Some are even recyclable.

Aluminum instead of steel. Aluminum is also being used in more manufacturing applications. In the 10 years beginning in 1976, for example, its use increased by about 5 percent a year. And a few years ago, Audi and Alcoa developed an all-aluminum prototype that weighed 46.8 percent less than conventional steel bodies. Lighter vehicles are more fuel-efficient. And aluminum is more impervious to corrosion than steel is.

Now, however, the use of aluminum sheet metal in the manufacturing process is beginning to decline as the use of composites and plastics climbs. But aluminum is still popular in a number of applications, such as wheels, bumpers, and under-the-hood castings.

Other manufacturing procedures and coatings. Chemical baths have been common for many years. They strip automobile assemblies of dirt, grease and oil so primers and paints adhere better. In addition, electrically charged primers assure even coverage and greater adherence.

Still in its early stages is the development of special paints — for example, paint containing "controlled-release" or water-soluble glass. When zinc alloys or similar ingredients are added to such paint, they are released slowly and consistently as the water-soluble glass dissolves.

SHORTCOMINGS

Despite all the advances in automobile design, today's cars remain vulnerable to rust. For one thing, to accommodate galvanization, thinner and thinner sheets of steel are being used. For another, pollution and acid rain are increasing. Moreover, each winter, more salt is dumped than the winter before. Other factors also affect a car's corrosion resistance.

Unibodies. Nowadays cars roll out of the plant in one piece called a "unitary structure" or "unibody." Instead of building the body around a separate frame, the manufacturers shape and then weld or bond the parts together in such a way that a frame is no longer needed. Each car ends up with a frame and body all in one.

The unibody is much faster and more fuel-efficient because it is lighter. But, as you might imagine, many of the box sections resulting from this type of construction are completely enclosed and inaccessible. If moisture works its way in — as it often does — it will probably go unnoticed until it has done extensive damage.

Welding. Although it makes for strong joints, welding alters the chemical composition of a car's protective coatings, including galvanization. It often leads to early corrosion. And, while adhesives are being used more extensively in car construction, welding is still the primary assembly method.

The new paint finishes. Most cars coming off the assembly line today are painted with a base coat (the color coat) and sealed with a clear coat for extra protection. These finishes prove to be very resilient. However, once they are damaged, spot repair can be impossible. Many cars must have entire panels repainted, even when the damage covers only a few inches.

So beware. Today's cars are more rust-resistant than ever before and will continue to improve, at least in the foreseeable future. But they still have many weaknesses. All it takes is one small imperfection for rust to invade the surface or structure of a car and start to spread.

Chapter 3
Preventive Maintenance

By inspecting your car at regular intervals, cleaning it thoroughly and frequently, keeping the paint finish protected and free of nicks and chips, eliminating leaks, and storing the car properly, you can ensure it a long and healthy life. In this way, you fight rust by making sure it never develops — an approach that can save you a lot of time and money.

To preserve your car, you must know its weak spots. You should also know what fosters rust and, conversely, what discourages it. Then you should plan your maintenance accordingly. If you are conscientious about it, you may never have to spend a dime on rust.

THE MOST DANGEROUS SEASON

One of the finer points of understanding rust is knowing when it is fiercest, and it isn't in winter — it's in the spring! Some experts estimate that corrosion increases by 20-40 percent in spring alone!

When water freezes — becoming ice, of course — it's dry, not moist. So it can't trigger corrosion. In spring, however, when that ice begins to melt, all the ingredients of rust are present: iron, oxygen and moisture. And in spring, melting snow becomes salt-laden slush, which gets thrown up against the car's underbody and into crevices or out-of-the-way places like wheel wells. Besides that, warmer temperatures and lots of slush create higher humidity, one of your car's worst enemies.

Knowing this, you can understand why timing can be everything. For instance, washing a car in the middle of January, when the temperature is hovering at 25 degrees and isn't likely to rise for days to come, will not be nearly so effective as washing that same car at the end of February, as soon as the temperature rises above 30 degrees, and when brief thaws can be expected over the next few weeks. (Of course, timing will vary from region to region.)

In such a case, washing your car in the middle of January *and* at the end of February (and all the weeks in between) would be the best option of all. See the box on page 41 for guidelines on timing washes, waxes, etc.

GETTING TO THE BOTTOM OF RUST

As I pointed out earlier, the bottom of your car sustains more damage more frequently than almost any other area. The reason is simple: it's nearest the ground. Mud, slush, pebbles, debris — even small rocks — get kicked up against it by the tires. The underbody also houses those confounding box sections and a host of cracks and crevices that trap and hold moisture. For this reason, maintenance should not only include regular inspections of the underbody — it should center on such inspections.

Don't confuse undercoating with rustproofing, either. Factory undercoating is usually a thick layer of sound-deadening material, not a process to discourage rust. When undercoating begins to wear, it flakes and sags and loses adhesion. Then dirt and moisture get trapped next to the metal, encouraging rust.

There are several ways to keep rust from infecting the underbody of your car. I recommend relentlessly thorough and regular cleaning, particularly in winter and spring.

Whenever you wash your car, try to tackle the underbody, too. If possible, raise the car on jack stands. When a car is positioned on jack stands, the wheels drop down and make it easier to clean the wheel wells. If you don't own jack stands, you can use metal ramps instead. Then, you will at least have room to move around under the car.

Ramps are often sold at reduced price at big discount stores. But be sure the rating for the ramps exceeds your car's weight. Remember, some cars with front-wheel drive have 60 percent or more of their weight over the front wheels. After use, rinse the ramps to remove all salt and mud and then store them dry to protect them from rust.

After your car is properly positioned, spray the underbody thoroughly, using a power nozzle or other attachment. Direct the water into all nooks and crannies and the wheel wells. Wash off all mud and salt deposits completely. If you don't, the water will only reactivate the corrosion process.

Have someone else do it. You may not want to wash the underbody yourself. It takes a lot of time and energy. It's also messy. Some carwashes have lifts and will steam-clean an underbody for $20-40 (taking approximately half an hour to do the job). Check your yellow pages under "automotive steam cleaning," or ask a fleet maintenance manager where he or she has vehicles steam-cleaned.

The best steam cleaners (the ones worth $20 or more) put the car on lifts. Some carwashes, on the other hand, charge $1 to $2 extra to drive your car over a stream of water that shoots up at the underbody. This is the least effective way to clean your underbody, since the wheels and suspension units block much of the water. But it's better than nothing. It does clean the floors and rocker panels, unless there are stubborn deposits clinging to them.

Inspect the underbody. When you finish washing the underbody, inspect it closely. Do this every six weeks in the winter and spring, less frequently the rest of the year (see the chart on p. 41). Check everything, even the exhaust system.

Clear any blocked drain holes and peel or scrape off (completely) any flaking undercoating.

Then treat the underbody with aerosol rustproofing (available in auto parts stores in several brands). You can also use engine oil or a special chemical (see Chapter 6). Just make sure you coat the entire underside. Any areas you miss will trigger an electrochemical reaction and may actually speed corrosion.

After the underbody has been thoroughly coated, check it regularly. Twice a year is a good interval (once in autumn and once in spring). Re-treat areas where the rustproofing has worn away.

Coating the underbody. Below are the steps to follow to coat your car's underbody:

1. Clean the bottom of the car with a power nozzle or high-pressure stream of water. Better yet, have it professionally steam-cleaned.

2. Clear any clogged drain holes, and go over the entire bottom with a plastic scraper (the kind used to remove windshield ice) or similar tool to keep from scratching good paint under loose undercoating. Then hose down the underbody again to rinse off any flakes of metal or undercoating you have dislodged. (See Chapter 6 for a list of chemicals for treating rusted areas.)

3. If there is any rust or scale on the underbody, treat it with a special chemical, or remove it completely and then prime and paint.

4. Spray on the rustproofing generously, working it into cracks and tight places. If you plan to coat the underbody with engine oil or a special chemical, cover the ground with newspapers or a roll of plastic. Apply the oil with a brush, in thick, heavy coats, working the bristles into all the cracks and

crevices. (If you are applying a special chemical, follow the directions on the label, or see Chapter 6.)

Don't coat the exhaust system, which heats to high temperatures and will burn the spray or oil off in no time. Do not coat moving parts, either. A detailed explanation of how to rustproof the interiors of box sections is given later in this chapter.

WASHING YOUR CAR

If I had to choose one preventive tactic and forget all the rest, I would choose regular washing. No other single measure does as much to prevent rust. Nor are there many other tasks that are cheaper or easier to carry out.

Nevertheless, there is a method to effective cleaning. If you are careless or in a hurry, you are just wasting your time. Make sure you have the proper equipment, the right place, and the time and energy to do the job correctly.

First, the experts say, never wash your car in direct sunlight. Ultraviolet rays, magnified through water beads, could cause the paint to spot. Second, never use a power nozzle on the body of your car — only under the front and rear bumpers, in the wheel wells, or beneath the car, as I explained above. A low-pressure stream of water is best for the other areas.

I use a special device that attaches to the water hose and shoots water at a higher-than-normal pressure. But I turn the water on low at the faucet and spray the car from a distance of several feet to reduce the pressure. This device also sprays soap or clear water (with the flip of a switch). I use this attachment to keep from having to stick my hand in a bucket of water. It is also faster, which is important in the winter.

Nevertheless, many experts advise against the use of any attachment at all. They recommend a soft sponge or terry mitt (with extra long nap), but never a brush. Bristles cause fine scratches across the surface. If your car is especially dirty, you may want to use a separate sponge or mitt for the worst parts. Also have on hand a large chamois cloth for drying.

Finally, use only detergents formulated especially for cars — not dish-washing detergents, general-purpose household cleaners, etc. But be careful. Many polishing compounds and waxes refer to themselves as "cleaners." Make sure you are buying only detergent, and that it is nonabrasive — or it could remove a little of your car's paint each time you use it.

If you aren't sure whether a product is abrasive, rub some between your fingers if possible. If it feels the least bit gritty, it's abrasive, and you shouldn't buy it.

Many experts believe you can get a car just as clean using no soap at all. After all, they say, detergents remove not only dirt, but any oils as well, and can weaken your car's natural resilience.

When you have everything you need to start the job, follow the steps outlined above for cleaning the underbody. Use a power nozzle, but set it aside before washing the rest of the car. (Of course, there will be times when you can't perform a complete underbody cleaning job. On such occasions, washing the rest of the car is preferable to doing nothing at all.)

Hose down the surface of the car thoroughly and let it sit for three or four minutes. This will loosen dirt and grime and make cleaning easier. (Warm water — never hot — is preferable to cold water, because it is better at dissolving salts and other chemicals lodged in the crevices of your car.) Hose down the surface again, wet your sponge and, starting with the roof, work your way to the bottom, rubbing lightly and wetting the surface frequently.

Older cars may have chrome-plated trim or other metal parts attached to the steel. Clean them the same way you clean the rest of the car: gently, using a specially formulated cleaner or no soap at all, and with a sponge or mitt — not a brush. Never scrape or attempt to brush dirt from these other parts. Chrome plating consists of ultra-thin layers. If you break the surface at all, corrosion is a sure bet. (Stainless steel, however, is relatively impervious to rust, even if it is scrubbed with a brush.)

You might consider removing the trim altogether to clean it. I removed the trim from just above my rocker panels when I painted my car. When I did, I discovered that at least half a dozen holes had been drilled to secure this strip. That's six places moisture, salt and mud can work their way in and trigger corrosion.

If you have one of the newer cars, the trim may be made of a composite material. If so, use a soft toothbrush or similar item to clean tight spots.

Don't forget to clean the door mounting areas. Moisture and dirt deposits often work their way in. Open the door and carefully sponge down these painted strips. Also clean the seam of the door skin, where it is folded on the inside of the door.

After you go over the entire body with the sponge and water, rinse the car thoroughly one more time. Wet a large chamois cloth and wring it out, and begin the drying process. Every few minutes, wet the chamois and wring it out again (a dry cloth can scratch the surface).

Once the car is dry, go over the surface, checking for tar spots, sap, or bird droppings. If you find any, use one of the special solvents developed for these stains (make sure it is compatible with your type of finish). You can find the solvents at most auto parts stores.

Now it's time to inspect the paint for nicks, chips and fine scratches and, if necessary, to repair them (see Chapter 4). Also inspect the following areas: door bottoms, exterior and interior; the battery box tray and accessories; the headlight, taillight and turn-signal housings, including the mounting screws; the floors; the bottom of the trunk; and around the wheels.

After that comes the wax job, and possibly polish or rubbing compound as well.

Automatic carwashes. Drive-through carwashes are usually better than no wash at all, but not always. Never take your car through one equipped with brushes; you'll wind up with a scratched surface. (Brushless carwashes generally have a sign that says they are brushless.) And don't go to a carwash that uses recycled water, which often contains remnants of salt and other chemicals. And it is probably best not to get in line behind an especially dirty car, as some of its filth may linger in the machinery and get dragged across your car, leaving scratches.

Look for carwashes that offer professional steam cleaning, particularly of the underbody. They are well worth the cost once or twice a year. Other carwashes offer bottom washes. While these are better than nothing, they are not as effective as steam cleaning or do-it-yourself underbody scrubbing.

RUB, POLISH AND WAX

Once your car is clean, seal its surface with a protective layer of wax. Wax fills scratches and cracks and keeps moisture and dirt out. A lot of people think wax makes a car shine, but it doesn't. The smoothness of the car's surface and the quality of the paint coating also contribute to its shine.

What wax can do by itself, however, is deepen the shine. A shine has depth when the color is rich (except, of course, for cars that are white, gold or silver) and when it reflects objects sharply. "Gloss" refers to the sheen of a car. On a dark car, depth is more important than gloss. The reverse is true for white, silver or gold cars.

If your car is new, regular washes and waxes will do for the first year or two. But as the car begins to age, the color will dull and the finish may appear cloudy or chalky. This is oxidation. It occurs as the sun's ultraviolet rays break down the chemicals in paint. The paint becomes more porous and soaks up oils, refusing to hold an even layer of wax. Yellow and red cars fade the fastest, blue and black the slowest.

When oxidation sets in, it's time to buy a good polish and apply it after the wash. But nowadays buying even a simple wax can be difficult. Product labels provide little information, and very different types of products are sold under very similar names. To ease some of the confusion, I describe the different types of polishes, rubbing compounds and waxes on the market today.

"Cleaners" or polishes. These are fine to medium-fine abrasives that are applied to the paint to rub out faint scratches and remove moderate stains. They should be used only on cars that are lightly oxidized. All polishes are abrasive, regardless of what the label claims. A product can't be a polish *and* nonabrasive. To do its job, a polish must remove a little of the outer paint surface.

One-step waxes. These are products that combine a light polish and a wax in one container. As you apply a one-step "wax," you remove a very small amount of the outer paint surface and leave a light residue of wax, usually not enough to be very effective. If possible, avoid these products.

Two-step cleaners or waxes. These products come in two packages. The first contains the polish, usually a fine to medium-fine abrasive. The second contains the wax, which itself may contain a very fine abrasive. After the car is washed, the polishing compound is applied and then the wax. I don't recommend two-step waxes either.

Polishing compounds. Polishing compounds are abrasives, usually of moderate coarseness — which distinguishes them from simple "polishes" (which are usually fine or very fine abrasives). These compounds are applied after the wash when a car is moderately oxidized, or when a lighter polish has not eliminated the chalky appearance. They should be followed with a very fine polish and then wax.

Rubbing compounds are strong abrasives, meant to be used on cars with heavily oxidized finishes or lots of cracks and scratches. They remove large quantities of paint; often they cut completely through the paint, exposing the primer. For this reason, they should be used only as a last resort and never with a machine.

Waxes. If a product is labeled as simply a "wax," it should contain no abrasives. But in the real world, this is rarely the case. Most waxes on the market today contain fine to medium-fine polish as well. Look for brands that claim to be pure wax or 100 percent carnauba wax. If you apply a wax with a clean white cloth and find that it begins to pick up some of the color from your car after only a few strokes, then that wax is abrasive.

Be prepared to try a few different brands. Try to buy those that offer full refunds and then, if you aren't satisfied with the job they do or if they turn out to be abrasive, get your money back. In addition, buy paste or liquid waxes, not sprays.

There are three basic types of nonabrasive waxes available today. The first is carnauba wax, derived from the plant. Carnauba wax usually produces the best shine and contains the most real wax. However, it tends to wear the quickest. It may also leave a cloudy or hazy area on the hood of a car the first few days after it is applied. That's because excessive engine heat causes many of the oils to migrate to the surface. Time and a couple of washes will solve this problem, however.

Polymers, the second category of wax, are not waxes at all but very durable clear protective coatings. They may last as long as two years. Although they leave a shiny finish, it doesn't compare to the depth produced by other types of wax. Moreover, some polymers may yellow. It may also be more difficult to match paint on cars with polymer coatings.

Some experts recommend applying pure carnauba wax on top of polymers, for a deep, durable, high-gloss shine.

Waxes with a silicone base are some of the most durable available, though they aren't as shiny as their carnauba counterparts and, unfortunately, may also contain abrasives. Most products on the market today combine two or more different types of wax — usually silicones and carnauba.

Polishes and harsher compounds can damage your paint. Before you use any type of abrasive on your car, you should know that it can rub completely through the paint if it isn't applied properly. Rubbing compounds are the most dangerous in this regard, but polishes can also cause damage. A one-step wax may rub through faster than a two-step product because it is designed to act fast. Never polish or apply a rubbing compound when you are in a hurry. And never use a machine to apply a rubbing compound — always apply it by hand.

If you use a machine to apply *polish*, make sure you use it properly. This means holding the applicator pad flat against the surface of your car, never at an angle. Switch to hand applications around door handles and other trim, and along sharp curves or raised body lines.

Furthermore, if you aren't sure which type of abrasive to use on your car, start with the finest grade you can find. You can always switch to a coarser grade if it doesn't do the job. Once you use a heavy-duty compound and remove large amounts of paint, however, the only way to repair the damage is to repaint.

In addition, over the life of your car, it's best to use as little abrasive as you can manage. Mix and match, if necessary. Use a very fine abrasive on areas of the car that are lightly oxidized and a coarser grade for areas in worse condition. See the box on page 24 for some guidelines.

Polishes and rubbing compounds do not protect your car. They merely smooth the surface, yielding a higher gloss and

---

KNOW YOUR PRODUCT

"Cleaners" and light polishes — For use on lightly oxidized surfaces, usually newer cars, to remove faint stains and light scratches. Use them no more than once a year, and always follow with a thorough wax application. In addition, always use a light polish after you apply a coarser abrasive.

Polishing compounds are medium-fine abrasives for cars with light to moderate oxidation and fine scratches. Use only when a lighter polish can't do the job. And once you apply a polishing compound (no more than once a year), follow with the finest-grade polish you can buy. Then seal the surface with wax.

Rubbing compounds contain coarse abrasives and should be used only as a last resort when a car has a heavily oxidized surface. More than one application over the life of the car is usually never called for. Other uses include blending paint around small repair jobs and eliminating surface imperfections on newly painted cars. Rubbing compounds should always be followed with a fine abrasive and then wax.

Wax — A nonabrasive final coating. Wax should be applied every three months, if possible. Wax fills fine scratches and coats the surface, keeping dirt, salt and moisture out.

---

richer color. Always apply wax after a polish or rubbing compound.

Applying polishes and rubbing compounds. Always start with a clean car. If you don't, you will rub dirt into the surface, leaving scratches. Once the car is clean and dry, park it in shade, out of direct sunlight. Take a clean piece of terry cloth or a

Turkish towel and fold it until it is 4-6 inches square and several thicknesses. If the product label specifies a different type of applicator pad, use that instead. The label will also tell you how much of the product to apply at a time.

Begin with a small section of the car. Apply the polish or compound in *straight lines only*, following the air line (the path air follows as a car accelerates down the road). Never use a circular motion to apply an abrasive. To remove as little paint as possible, use an applicator that is slightly damp and apply less pressure as the compound dries out. Do not get any compound on the windows. The label will usually specify other areas to avoid as well.

After you apply compound to the entire car, hose it down. This will remove any fine particles of polish. Rinse the car thoroughly. Any polish left in crevices or cracks will attract and hold moisture, causing rust.

Metallic finishes. If your car has a metallic finish, be extremely careful. Never use more than a very fine abrasive, and apply it with as little pressure as you can manage. When you buy any product, make sure it is safe for use on metallic finishes. You might want to ask your dealer what products he or she recommends, or check with a salesperson at a reputable auto parts store.

Base-coat/clear-coat finishes. As I explained in Chapter 2, new cars come off the assembly line with a base coat (the color coat) and a clear, protective coating over that. Don't use heavy abrasives, like rubbing compounds, on these finishes. Instead, use a detergent, wax or fine abrasive specially formulated for them. Ask your new-car dealer to recommend a brand, or inquire at your local auto parts store. (Some carmakers sell car-care products specially formulated for their line of cars. These products are usually more expensive than similar products in auto parts stores.)

Applying wax. Use a clean cloth to apply wax, preferably a piece of terry cloth folded to several thicknesses. Turkish towels are also good applicators. Start in the middle of a panel and work your way out, using a circular motion. It's probably best to do the sides of a car first, when you have the most energy, and to end with the horizontal surfaces.

Apply the wax conservatively. More wax is not better. In fact, if you apply too much wax, you may end up rubbing more off than you leave on. Rub the wax on a small area — say 18 inches square — and then go back with a clean dry cloth and buff it until it shines.

You may choose to machine buff the surface. If so, use the proper pad and follow the directions outlined above for applying polish with a machine.

A newly waxed surface repels water well, resulting in almost spherical beads. Your car needs a new application when water puddles on its surface instead of beading. When the beads get to be more than 1 inch in diameter and are low to the surface, wax again.

Paint afflictions. Paint jobs can develop a host of problems if they aren't properly maintained. And different treatments are necessary for the various ailments. To help you diagnose your particular problem, I include a brief description of some of the most common paint disorders.

Spotting consists of a series of small rings where water has evaporated from the surface. Spotting usually occurs on older or badly weathered finishes. Try polishing with a fine abrasive.

Color spots and freckles appear when your car is exposed to soot and other airborne pollutants or acid rain. This discoloration can be permanent. Try washing and waxing more frequently to prevent further damage.

Crazing is a disease peculiar to cars with metallic finishes. It develops when the metal-flake coating begins to contract and crack. When wax or other debris gets caught in these cracks, the surface looks streaked. You can help prevent crazing by keeping your car out of direct sunlight and other harsh elements. To treat it, use a very fine abrasive specially formulated for metallic finishes, and apply it carefully. If that doesn't help, the problem is probably too advanced for you to handle. Consult a professional paint shop or auto detailer.

Etching results when acids, tar or other elements are allowed to sit on the surface for an extended amount of time and the paint gets eaten away. Clean your car frequently and try not to park under trees (to avoid bird droppings and sap).

Below are some more tips on polishing, waxing and applying rubbing compound to your car:

- Waxing a hot surface may result in streaks. A streaked surface may also indicate that you did an inadequate job of applying the polishing or rubbing compound, or that you simply you didn't use enough wax.

- Never use an abrasive not specially formulated for cars.

- Be extra cautious with metallic finishes and with base-coat/clear-coat finishes. Make sure any product you buy is compatible with your finish, and never use a rubbing compound on a metal-flake paint job.

- Never use an abrasive on a car less than 12 weeks old or you may ruin the paint coating (it takes 12 weeks for paint to cure). Keep a brand new or newly painted car clean and out of the weather.

- Always follow a medium or coarse abrasive with a fine-grade polish and then wax.

- Always apply wax or abrasives with clean cloths or pads.

- Apply polishes and rubbing compounds in straight lines only. If you use a circular motion, your finish will look wavy afterward.

- Always read product directions.

PREVENT, PINPOINT AND REPAIR LEAKS

Body leaks are one of the worst perpetrators of rust. You can wash your car thoroughly several times a week, take meticulous care of the underbody and paint surface, and still end up with rust holes caused by water leaking into the car. So a good prevention program should include regular inspections of the car's weatherstripping and punctual repairs of leaks.

Leaks are most common around doors, windows, windshields and trunk openings; around the cowl; around sun-roofs, T-tops and on convertibles where the top and car body meet; under floor mats; around headlights, taillights and turn signals; around body molding; and along trunk bottoms.

Most of these leaks are the direct result of faulty weatherstripping or gaskets. Determine the extent of the leaky weatherstripping. Then reglue it with a special adhesive or repair it with silicone sealer. If necessary, replace it completely.

A good way to test for leaks is to run water over the area you suspect. For example, if the leak appears to originate in the upper corner of the windshield on the passenger side, sit in the passenger seat while a friend floods the glass and the metal above it with water. Watch the area for a trickle of water or any moisture that works its way through to the interior of the car. Once you pinpoint the source of the leak, clean and dry the area thoroughly.

Treat any rust you find, and spot-paint the area, if necessary (see Chapter 4 for guidelines on rust repair). Then apply a small amount of silicone sealant, covering the area of the leak and extending the application a little to each side to ensure a com-

PROTECT TAILLIGHTS & TURN SIGNALS

- Remove the lights and disassemble them.

- Inspect the gasket — the rubber seal around the edge of each assembly. Is it intact? Or is it worn and cracked? If it's damaged, repair it with silicone sealant or a special adhesive. Or remove it completely and install a new one, making sure it's a perfect fit.

- Check the housing, the part of the assembly that holds the bulb. Is it corroded? If it is, scrub it with a wire brush to eliminate rust. Remove any rust you find in the socket as well.

- If the housing is corroded, have it replated. If it's plastic, it can be painted or lined with aluminum foil, with the shiny side toward the lens. Use an adhesive to hold the foil in place.

- Spray a bit of lubricant in the socket before reinserting the bulb. This will discourage further rust.

- Reassemble the lights. Is there a watertight seal between the housing, the gasket and the body of the car? There should be.

- If you repaired the gasket with a sealant, allow it to cure for 24 hours or more, then test for leaks.

- At the boot — the rubber fixture where the electrical wire enters the light assembly — apply heat-shrink tubing to the joint between wire and rubber to keep moisture out. The putty used on electrical cables that pass into buildings is an alternative. It's sometimes called connector or cable sealant.

plete seal. Let the sealant cure 24 hours and then conduct the test again.

If water is leaking between the body of the car and the rubber gasket around the windshield, insert a fine-blade knife about one quarter of an inch under the gasket. Move the knife slowly around the windshield, applying silicone between the gasket and the metal.

If you suspect that water is leaking in around the trunk lid, you may have to lie inside the trunk with a light and watch for drips or trickles as someone else floods the rear of the car with water. Be a detective. Leaks almost always leave tracks or stains.

Small rust holes in the floor of your car or trunk will become big rust holes in no time if you don't repair them, thanks to water or slush on the street. To check for such holes, put the car on jack stands, remove the wheels, and spray a heavy stream of water from underneath the car. Then look to see whether any got through (it may be necessary to remove the carpeting). If it did, repair the holes immediately (see Chapter 4).

Sometimes the floors may be wet, even when they appear to be dry. That's because some floors are covered with sound-deadening tar mats that are painted the same color as the floor and other interior areas. As a result, they appear to be metal. Scrape them gently with a screwdriver. If you find tar, it's best to remove it. I have removed tar from "dry" floors only to discover one-fourth cup of water rotting the metal underneath.

This tar covering can be softened with some chemical paint strippers or a heat gun. Be careful if you use a heat gun. You could start a fire.

Checking weatherstripping. To check the general effectiveness of your weatherstripping, conduct this test. Insert a dollar bill between the top of the door (or edge of the trunk or window) and the body of the car. With the door closed, pull the dollar bill free. You should feel at least some drag. If the bill comes out too easily, then your weatherstripping is not working properly (or your door is hung incorrectly) and should be repaired or replaced.

To repair it, build up worn or missing spots with silicone sealant. Use an automotive weatherstrip adhesive to reattach portions that have worked loose. Torn weatherstripping can hold water like a sponge, causing the area to which it is attached to rust.

If the weatherstripping has to be replaced, buy a new piece from a car dealer or auto parts store. Or order duplicate original-equipment weatherstripping in bulk quantities from mail-order companies (see Appendix 1). But make sure it is made specifically for the area in which you plan to install it. And be sure to remove the old weatherstrip completely. Pry it up with a screwdriver and use a gasket-scraping tool to remove any excess, but don't scratch the surrounding metal. Then use a screwdriver or similar tool to force the new weatherstrip into the groove in the door or window. Finish sealing the area — if need be — with adhesive specially formulated for weatherstripping.

Obviously, any rust, scratches or peeling paint should be repaired and repainted before you install the new weatherstrip. Allow the paint to dry thoroughly before you use any adhesive on it. Some adhesives can soften paint, causing it to flake.

If your weatherstripping is in good condition, keep it that way by spraying it regularly with silicone spray.

Clearing drain holes. Every car comes equipped with drain holes in the door bottoms, to let moisture work its way out. Many cars also have drain holes called "darts" (because they are shaped rather like the head of a dart) in rocker panels. In addition, many cars have drains at the bottom of the heater housing, in the cowl or air intake at the base of the windshield, and in the trunk.

Examine your car thoroughly and pinpoint all the drain holes. Then, when you inspect the underbody, make it a point to clear them. Use a pipe cleaner to probe. But don't scratch the paint surface, or rust may develop.

If you rustproof your car — or pay to have someone else do the job — check the drain holes about two weeks later. Rustproofing solutions often filter down and clog drainage. Finally, after you wash your car, leave the doors open a few minutes so excess water can drain out.

Some cars come from the factory with plugs in their drain holes. This is especially true of cars that are shipped from abroad. These plugs keep sea air from entering enclosed body panels. The dealer is supposed to remove them before the cars are sold. Many forget to do so, however. When I removed these plugs from the rear quarter panels of one car, a quart of water drained out!

If you find rust beginning to develop, you may want to treat the area with a neutralizing chemical (see Chapter 6). But save the plugs. Some rusted areas must be kept damp with the chemicals; the plugs will keep them from draining out.

PROTECTIVE DEVICES

There are several accessories designed to protect your car from rust or from dents and scratches that can lead to rust. Among them are urethane strips, bed liners for trucks, electronic devices designed to derail the corrosion process, and plastic or vinyl body side moldings.

Protective moldings. These are long strips of plastic (or composite materials) or vinyl that protrude slightly from the side of your car. They are especially useful in crowded parking lots, where people often open their doors into the sides of other cars.

On many of the new cars, body side moldings are standard equipment. They can also be ordered as a special option or as

---

PROTECT YOUR CAR FROM LEAKS

- Keep floor and trunk mats dry.

- After washing your car, leave the doors open for several minutes so excess moisture can drain.

- Periodically clear the drains in door bottoms, rocker panels, the cowl and air intake vents at the base of the windshield, and at the bottom of the heater housing.

- Treat weatherstripping regularly with silicone to prevent cracking and dryness.

- Always remove old weatherstripping completely, and clean the metal surface thoroughly, before installing a new strip.

---

part of a package of several different options. If you are ordering such an option (or options), make sure the moldings are protective, not purely ornamental. (See Chapter 8 for more specifics on buying a new car and selecting options.)

I have also seen door guards made from ensolite. I don't know where you can buy these, but you can make them yourself by attaching magnets to strips of ensolite using duct tape. The magnets will hold the strips against the side of your car. Ensolite is available in camping supply stores.

Urethane strips. "Urethane film" is a transparent, adhesive-backed strip that adheres to the surface of your car and protects it from pitting. It is usually recommended for rocker panels and wheel wells, since these areas are exposed to pebbles and other debris kicked up by the tires. These areas are also more difficult to keep clean.

To install a urethane strip, tape the film to your car and trace the outline of the area you want protected. Then mix one part isopropyl alcohol and three parts water, and spray this solution on the application area. Peel a small amount of the urethane backing off and attach an edge of the film to the car. Then slowly peel the rest of the backing off. Once the strip is correctly positioned, press out excess moisture and any bubbles that have formed.

There are also protective coatings that can be mixed with paint and applied to areas that need extra protection, such as rocker panels. Some transparent aerosol sprays incorporate this extra protection as well.

Bed liners. Truck beds are especially susceptible to corrosion because they suffer so much wear and tear. Lumber, bricks, furniture and a host of other materials get thrown in the back of a truck at one time or another.

Bed liners are rubber mats of varying thickness that fit across the surface of the bed and supposedly protect it from chips and scratches — and thus from rust. But they can also trap moisture between the rubber and the metal, thereby *encouraging* rust. For this reason, it is best to install a liner only on a newly painted truck bed — one with a healthy, unscratched surface. That way any moisture that gets caught under the mat does little damage.

Also make sure the liner fits snugly and securely and gets anchored with a minimum of screws (which require the drilling of holes). And treat any holes that are drilled to secure the lining with a rustproofing chemical right away. Finally, remove the liner periodically to inspect for rust.

Electronic cathodic devices. As I explained earlier, rust is an electrochemical reaction. As such, it involves an anode, or positive charge, and a cathode, or negative charge. When an electrolyte (such as moisture) is present, the difference in charges between the anode and the cathode causes the metal to dissolve into the electrolyte. As it dissolves, it mixes with oxygen. This is oxidation.

When bare metal is exposed on a car, it acts as an anode and sacrifices itself in place of the negatively charged cathode. In other words, it dissolves, while the cathode — the painted surface — remains intact. Electronic cathodic devices circumvent this process using DC current from your car's battery. They require the installation of one or more control modules and two or more anodes.

Rust Evader (the only such device being marketed in the U.S.) supplies current to the metal car. The car then becomes a capacitor. It begins to store and discharge electrons simultaneously. What distinguishes this unit from classic cathodic devices is that it shuts the current off periodically, then switches it back on, causing electron discharge. This discharge occurs at scratches, paint chips and other common sites of corrosion. As a result, the metal is preserved (not to mention the battery, which might run down otherwise).

These devices work better on cars that were manufactured after 1976. That's because the carmakers began coating many interior surfaces with primer in 1977, and have continued to do so. When there is a lot of bare metal on the interior of a car — as there is on older cars — the cathodic devices must work overtime to protect the metal from corrosion. Even then, in many cases, corrosion may strike.

In my opinion, these devices will be more effective on an older car if its interior panels are coated with penetrating oil (see Chapter 6), or if the car is chemically stripped (using the immersion process) and its interior surfaces coated with primer and paint. That's because these cathodic devices are much more efficient if there are fewer corrosion sites requiring electrons. By coating interior surfaces with oil or primer and paint, these sites are protected from oxygen (and the electrolyte). In other words, they are no longer corrosion sites.

The manufacturer of these devices says only one set of conditions could possibly limit their effectiveness: when a pocket of moisture against the metal is surrounded by a completely dry area. In such a case, the electrons would not be able to reach the pocket unless the environment were extremely humid. So it is still important to keep your car's underbody free of dirt and salt deposits.

If you own an old car with a 6-volt battery or a positive-ground system, or both, and the car is stored in a garage most of the time, a cathodic device can help. First, install the device. Then disconnect the 6-volt battery. Next, hook the device to a 12-volt battery in your garage, and connect the 12-volt battery to a trickle charger. This is better than connecting the cathodic device directly to the trickle charger. That's because the 12-volt battery provides *continuous* direct current. The trickle charger converts alternating current through a rectifier. As a result, its current is pulsed, or intermittent.

Rust Evader sells most of its devices directly to new-car dealers. But they are also available to the public through J.C. Whitney, an independent mail-order distributor. At a dealership, the cost for these devices may range from $395 to more than $1,000, depending on geographical location (but this price includes a very easy installation). Through J.C. Whitney, a two-anode device goes for about $150. (A four-anode device is also available but, in my opinion, two anodes are sufficient for passenger cars.)

Rust Evader offers a "limited lifetime warranty" to the original owner with no mileage limit for new cars. Used cars (less than three years old with less than 150,000 miles) can be covered up to their sixth year. Some important provisions are (1) The auto manufacturer's warranty is the car owner's first recourse; (2) repair costs may not exceed the National Automobile Dealers' Association (NADA) wholesale book value; (3) the car must be brought in for inspection every two years; (4) corrosion that is caused by battery acid or corrosion that affects the passenger floor is excluded; and (5) the car owner must notify the warrantor within 30 days after a possible claim is discovered.

While corrosion caused by battery acid may be an obvious exclusion, floor rot may not. But I have seen floors deteriorate quickly because of water leaking around the windshield or door gasket. There are other exclusions as well. If you install the unit yourself, the warranty may be even more limited.

Rust Evader works best on small patches of bare metal, such as cracks, seams and chipped paint. It does not work as well on large, bare areas. For example, it would not protect the center of a 6-inch circle of bare metal, but it would protect the perimeter.

Rust Evader Corporation also markets a protective coating called AEL. When used in conjunction with the cathodic device, it is warrantied to protect the paint from acid rain pitting. This coating must be reapplied every two years. For more information on Rust Evader, see Appendix 1.

PREPARING FOR WINTER

In late fall, before rain and snow set in, you should clean your car from top to bottom and prepare it for the months ahead. It is very important that you take time to clean the underbody thoroughly and that you inspect it afterward and remove any flaking or loose undercoating. If you can only get around to this complete cleaning once a year, at least time it properly: do it in the fall!

Eliminate any pockets of mud and sand; wash the car; clean sap, tar and bird droppings from the surface; and repair any rust spots or chips you encounter (see the next chapter for specifics on repair). Also apply rustproofing to the underbody, if necessary (or engine oil). In short, make sure every bit of exposed metal on your car has a protective coating, be it paint, rustproofing or oil (see Chapter 6). Then wax your car. If it needs a polishing job, this is the best time to do it. But make sure you seal the surface with a good nonabrasive wax.

Finally, check the trim and determine whether it is made of plastic/composites, chrome or chrome-plated steel, stainless steel or anodized aluminum. Wax (with a nonabrasive wax) aluminum and plastic/composites. Coat chrome, chrome-plated steel or stainless steel with engine oil, wiping off the excess.

Parking your car. Once winter sets in, you must park your car in the least corrosive environment possible. Most people assume that this is indoors, in a garage. And why not? After all, in a garage, it is sheltered against the damaging effects of acid rain or sudden hail — even against the effects of sulfuric acids, which lace the air in industrial regions. But I'm afraid the issue isn't quite so simple. There are several factors to consider when deciding where to keep your car from day to day. The first is how clean your car is. The second is the condition of your garage. The third is the weather.

If your car hasn't had a thorough cleaning in several weeks and there's a good chance there are salt or mud deposits underneath it or behind the wheels, and if your garage is heated or seldom goes below freezing, then don't park your car indoors. The reason — if not already apparent — is simple. In a heated garage, the moisture trapped in the salt and mud deposits will activate the corrosion process. Under such circumstances, you are better off parking your car in a driveway or along the curb if it's 32 degrees or below outdoors. Then the moisture trapped in the deposits will stay frozen and do little damage.

On the other hand, if your car is clean and deposit-free and your garage is cool and dry, then you are better off parking indoors. In fact, studies have shown that cars garaged under such conditions have a 25 percent greater chance of escaping rust.

Car covers. If you don't have a garage, you can probably get along fine parking on the street or in a driveway, as long as you monitor the weather and the condition of your car. But if you go for long stretches of time without driving your car, or if you live in a particularly harsh environment, you might consider buying a good car cover.

First, make sure any cover you buy is designed specifically for your make and model of car; a poor fit will cause more problems than it solves. Next, make sure you tie it down securely when you use it so the wind doesn't blow airborne debris inside and so small animals stay out.

Also make sure it allows your car to "breathe" — in other words, that it does not trap and hold moisture. Finally — and this is most important — never use a cover unless your car is absolutely clean and dry, from top to bottom.

Storing your car. If you aren't going to be using your car for several weeks or months, there are measures you can take to protect it from rust. First make sure it's clean and dry. Then do the following:

1. Check the surface for small rust spots and nicks or chips and repair them (see Chapter 4). Apply new rustproofing to the underbody. If you own an older car with no underbody rustproofing, and if your car is showing signs of corrosion, apply a coat of "bridge paint" to the underbody (see Chapter 6). Remember, however, that if you don't coat the entire underbody, the corrosion will probably worsen. And you won't know about it until it's too late!

2. Change the engine oil and filter, and fill your car all the way up with transmission fluid, oil, antifreeze or water, and gasoline (do not over-fill). Keeping your car full of these fluids reduces the risk of condensation in its inner workings. Condensation leads to rust.

3. Remove the spark plugs and fill each hole with a tablespoon or two of oil. Then turn the engine over a few times to distribute the oil properly. Reinsert the spark plugs.

4. For especially long storage periods, seal the intake and exhaust openings and lay newspapers over the engine to absorb moisture.

5. Remove the battery completely and store it separately. Hose out the battery tray, making sure you rinse away all traces of acid. Repair any damage.

6. Spray the weatherstripping with silicone and repair any defects.

7. Close the windows completely to keep small animals out, but leave the fresh-air vents open. Make sure the floor mats are clean and dry, and place some moisture-absorbing crystals (sold in hardware stores) in the interior of the car. Also check the trunk mats for moisture.

8. Apply an even layer of wax but don't buff it. Wax is more durable if it isn't buffed.

RUSTPROOFING YOUR OWN CAR

Rustproofing your own car is a big and time-consuming job. Rustproofing means spraying or brushing a rust inhibitor on all unpainted metal surfaces, including the interiors of box sections, doors, rocker panels, fenders, and so on.

It's a big job because, once you start, you can't stop until all unpainted metal is coated — or rust may develop even faster than it normally would. It's time-consuming because it involves a lot of preparatory work — such as cleaning, making minor repairs, and masking off the braking and exhaust systems, among other areas. It also requires a slow, thorough approach.

Some experts claim that a complete rustproofing job can add 10 or more years to the life of a car. It certainly can do a lot to

PREVENTIVE MAINTENANCE

Weekly

- Wash
- Check floor mats
- Check trunk mats

Monthly

- Check paint surface for nicks and chips
- Clear drain holes
- Wash underbody

3 Months

- Apply wax
- Inspect underbody
- Check weatherstripping and test for leaks

6 Months

- Steam-clean underbody
- Rustproof underbody where needed

Note: In winter, salt-belt cars should be washed twice weekly, if possible.

protect your car from rust, provided the job is properly performed. However, some manufacturers discourage car owners from applying rustproofing solutions to new cars (see Chapter 9). But if you rustproof a car, inspect it thoroughly every year or so to ensure that the chemicals are still working and that the metal is still completely coated.

Many professional rustproofers refuse to work on cars that are more than three months old. Some will work only on cars that are one year old or younger. The reason for this is pretty simple: the longer a car has been driven, the greater the network of fine cracks and crevices that trap and hold dirt, moisture and grease along the underbody and inner panels.

These dirty cracks and crevices are where rust develops first. So the rustproofers would rather not fool with an older car, where rust has probably already begun to spread.

Whatever the age of your car, the first step in rustproofing is a thorough cleaning job — the most thorough cleaning you can arrange. I recommend you have your car professionally steam-cleaned for the occasion. That will eliminate most of the dirt and grease, and any flaking metal or undercoating. High-pressure cleaning — a combination of water and high-pressure air — is an alternative that will certainly lift old, peeling undercoating. And it is best done on a lift.

But even after this extra cleaning, your car will still hold some dirt and grease. So put it on jack stands and inspect the underbody yourself, using a bright light, a hose with a small nozzle, and a long, blunt object.

While most body mechanics use screwdrivers to poke around for loose metal and hidden holes beneath the undercoating, I don't recommend it. I prefer some sort of blunt object. With a screwdriver, you could scratch the paint on good metal beneath the undercoating. This scratch will then start rust that will creep under the paint. A sturdy plastic ice scraper is excellent for finding loose undercoating.

Go over every section of the underbody, scraping or prying loose any grime or rusted metal you find. Also remove any loose undercoating. Then spray the area with water and clear any blocked drain holes you find (but remove the light first). Finally, remove the wheels and go over the inner wheel wells in the same manner (you will want to leave the wheels off throughout the rustproofing procedure).

You can get a higher-pressure stream of water to rinse the underbody and wheel wells of your car if you cut a 12- to 18-inch piece of 1/2-inch (or 15-millimeter) copper tubing and attach one end to your water hose with a clamp. Flatten the other end with a hammer, leaving a narrow slit for the water to pass through. This will shoot water into the cracks and crevices with greater force.

Rusted areas will have to be repaired before you can begin to rustproof. If you encounter a lot of rust beneath the car, you should have it repaired by a professional welder, especially if it involves a structural area.

For lighter rust problems, however, a rust neutralizer or converter (or other chemical) may do the trick (see Chapter 6). But first scrape and wash away any loose metal you find. Then apply the chemical, coating the area twice if necessary. When the chemical has taken effect, coat the area with two layers of primer and let it sit overnight at least, preferably longer. (But read the product label; some rust converters require no primer.)

Rustproofing is easier if you understand how your particular car is constructed. If you can locate a diagram of all the parts of your particular model in exploded detail, or a good cut-away drawing, you will be ahead of the game. Factory service manuals often have drawings of this kind. The owner's manual sometimes does. A local body shop or body supply store may let you look at their diagrams.

If you can't find such a diagram, examine your car yourself from top to bottom, noting the locations of box sections, welded seams, drain holes and other openings in the metal through which you can gain access to interior areas. Some box sections are open at one or both ends. In addition, you can often remove door striker plates, light switches, metal trim, and some screws and bolts to reach inner areas.

Sometimes rocker panels will have 1.5-inch to 2-inch holes stamped in them at intervals. These are usually hidden behind interior trim (carpeting, vinyl or plastic). You can use these holes to apply rustproofing to the interiors of the rocker panels.

You will probably also have to drill a few holes to gain access. If you do, coat the raw metal around the hole with two layers of primer and, once the primer is dry, with rustproofing solution or paint.

Rustproofing solutions come in a variety of packages. At least two companies offer home rustproofing kits, which include three to four cans of the rustproofing solution, several spray wands and nozzles, plugs to fill any holes you have to drill, and instructions.

The kits are available at most auto parts stores. They usually contain a solution for interior areas like box sections, and one for exterior areas, like the underbody in general. The solution can usually be sprayed or brushed on. Spraying is recommended for box sections and tight places, brushing for large, flat surfaces.

If you don't buy a kit, make sure the rustproofing solution you purchase is just that — rustproofing — and not undercoating or sound-deadening. Undercoating and sound-deadening formulas only add a layer of insulation to cut the noise between the passenger compartment and the road and mechanical systems. If you want to fight rust, don't use these products. Most of the time they just trap moisture and dirt against the car.

Also make sure the rustproofing solution you buy for interior areas is thin enough to creep into seams, flanged metal edges, and deep cracks and crevices, yet viscous enough to cling to vertical surfaces. You can always thin it down, of course. Just buy the appropriate solvent.

Before you begin rustproofing, mask off areas of the car that need to be protected from the solution. These include the engine and gearbox, the rear axle, any rotating parts, the drive shaft, the half shafts, the brake drums or discs, the brake cables, any door-mounted speakers (it may be easier to temporarily remove these), the exhaust system, lubrication points on the steering and suspension, suspension units, the hand brake, any painted areas in the line of spray, and the battery and battery tray.

Cover these areas with newspaper or light plastic and secure the covering with masking tape.

Put the car on jack stands (if it isn't already there) and remove the wheels. Also dress appropriately. Wear old clothes that cover your arms and legs, and a hat and gloves as well. Wear goggles and the appropriate mask when you apply the solution. And make sure the temperature in your work area is 50 degrees Fahrenheit or above.

Then gather the rustproofing solution, a paint brush for use on large flat areas, a good strong light, the spray wands and nozzles (or a length of plastic tubing), an electric drill (and a bit), a small container of primer and a small brush (for the primer), and plugs.

If you have to drill holes to gain access to the interior, try to drill them in vertical planes rather than horizontal ones because vertical sections undergo less torsional stress. Seal all the holes you drill.

Your biggest challenge will be coating interior areas, like box sections, thoroughly. To prepare for the challenge, build a model box section by folding cardboard into a box shape (with no open sides) and seal the seams with masking tape. Then make a small hole in one end of the box.

If you are using a rustproofing kit, then the solution is ready to spray. But if you purchased the chemicals separately, they will probably have to be thinned (read the label).

Rustproofing kits provide you with the proper wands and nozzles. But a number of containers and spray nozzles are available at auto parts stores and through the mail (see Appendix 1). In most cases, these attachments are superior to the ones provided in the rustproofing kits.

Once the model box section is completed and the rustproofing solution properly thinned, attach a nozzle that sprays 360 degrees to your spray wand or tubing. (You can make such a nozzle, if one doesn't accompany the kit you buy, by sealing off the end of a piece of plastic tubing with a self-tapping screw, and then piercing a ring of very small holes in the tubing just below the screw; see the diagram on p. 48.) Insert the nozzle in the hole in the box and extend it until it reaches the other end.

Then begin spraying the solution inside the box. Slowly withdraw the nozzle, spraying continuously. When you finish, cut the box open and see how much of the interior you managed to coat. Also note any variations in the coverage different parts of the box received.

Some minor adjustments in the nozzle itself, in the location of the entry hole, in the rate at which you withdraw the nozzle, or in the length of the tubing or wand you use, may be necessary.

Before you begin applying the rustproofing solution, cover the ground under the car with newspapers or a light plastic drip cloth.

Areas of focus. Some areas of your car are more prone to rust than others, and they should get extra attention during the rustproofing process. For example, metal that has been welded, hammered or cut away during rust repairs is weaker and more susceptible to corrosion than the metal in other parts of the car. So make sure the rustproofing solution covers welded seams and repair areas thoroughly.

On vintage autos that have some wooden parts, apply extra solution where wood and metal come together. Also remove all chrome strips and other metal trim and apply rustproofing solution behind them.

On the underbody and in wheel wells, spray (or brush on) an even coating of the exterior solution, applying it over any existing factory undercoating. Spray inside box sections with the interior formula.

For rear fender quarter panels, spray the interior formula. You may be able to gain access to these panels through the trunk. If you can't, you will have to drill an access hole and insert a length of tubing or a long wand, moving it around the area as you spray to ensure complete coverage.

For rocker panels and front and center door posts, you can often gain access by removing any trim or the door striker and courtesy light panels. Otherwise, drill a hole near the bottom of a post and spray from there.

Roll up the window before you spray inside a door. Also remove any speakers mounted in the door. On many cars, the interior door panel can be removed for greater access.

Spray the sheet metal on the underside of your hood with the exterior formula. Also coat any unpainted metal areas in the trunk, including the metal between the stress panels on the trunk lid.

YOUR CAR'S WORKING PARTS

Rust isn't limited to the body of your car. It strikes many operating systems as well, like the radiator and the brakes. Keep this in mind when you perform routine maintenance. Silt, sand, airborne debris and microorganisms can work their way into your car's working parts and accelerate rust. Below is an explanation of some of the basic safeguards you can take.

The cooling system. If you don't change the antifreeze in your cooling system very often, you may be contributing to

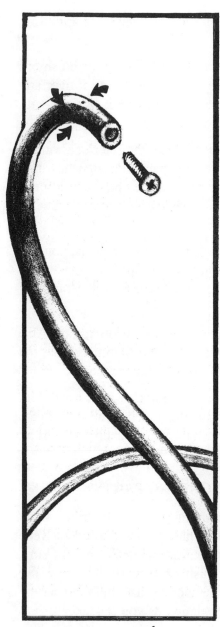

spray nozzle

corrosion. That's because antifreeze loses its rust-fighting ability after a while. Then corrosion or erosion (or both) can set in, and deposits may begin to build up in the radiator and heater passages.

Erosion may even afflict your car's cylinder head, if it is made of aluminum. So get in the habit of flushing out your cooling system annually. Then refill it with new — and potent — antifreeze.

Even if you live in an area where it never goes below freezing, it's a good idea to use a 50/50 mix of antifreeze and water to prevent rust. Antifreeze also raises the boiling point of water and helps prevent overheating.

Corrosion in the cooling system usually occurs first in crevices or tight places, like freeze plugs or hose necks. If you have excessive buildup in your cooling system, there are cleaners and flushes on the market, and available at most auto parts stores, designed to treat such problems. If you have an aluminum cylinder head, make sure the cleaner/flush you buy is compatible with it.

One automobile mail-order company offers an inexpensive device that releases zinc in the radiator, at the filler neck, to prevent rust in the cooling system (see Appendix 1).

The braking system. Because of the high temperatures it is subjected to and its precarious position on the body of the car, the braking system is particularly vulnerable to rust. On many cars, especially older cars, the braking system is simply an appendage to the body instead of an integral, protected component. As a result, brake lines, cables and connections are exposed to mud, slush and other road debris kicked up by the tires.

Because of this, frequent underbody cleaning and inspection are vital to the life of your braking system.

In addition, conventional brake fluids absorb moisture over time. So use silicone brake fluids. Or drain the brake system periodically (some manufacturers recommend two-year intervals) and refill it with fluid from a new, unopened can.

If you encounter rusty brake parts, a good aerosol penetrating solvent may help to free them. If disc-brake pistons stick or jam, the cause may be a corroded cylinder, in which case the entire cylinder assembly will probably have to be replaced. Check the piston seal from time to time as well. When this seal is intact, it helps lock out moisture and dirt.

The exhaust system. The exhaust system is also prey to high temperatures and corrosive elements and thus should be inspected regularly. Frequent cleaning helps, which is fairly straightforward under the car, but more difficult on exhaust parts under the hood.

If exhaust clamps or bolts rust completely through, you may have to cut them off using a hacksaw or, in the case of rusty nuts, bolts or studs, a chisel. They can also be removed with a torch. In many instances, it may be more cost-effective to let a muffler shop do the work.

The ignition system. Spark plug wires and distributor cap components often succumb to rust. When this happens, the wires must be replaced. However, you can usually clean the terminals and rotor in the distributor cap with fine sandpaper. Otherwise, replace the cap.

Sometimes, on older cars with electronic ignition systems, the terminals in the ignition module corrode and foil attempts to start your car. Clean them with a wire brush. However, on newer, computerized cars, these terminals appear to be corroded when they're not, so leave them alone. If you try to brush them you will just cause further damage.

Electrical connections. Waterproof electrical connections by sealing them with extruded rubber tape or heat-shrink tape. Simply cut it, mold it and apply it (or insert it, if you have plastic connectors).

Clean corroded electrical connections with a penetrating spray and a wire brush. If the terminal is locked in a plastic connector, you may be able to scrape it clean using a fine-tip screwdriver or miniature files. I have also used a Dremel tool with a 1/16-inch rotary file on the slowest speed.

Apply silicone sealant to the wire where it enters the connector. This keeps moisture and salt out. Finally, spray the connector with a thick waxy marine penetrating spray to prevent corrosion.

The fuel system. Moisture can build up in the gas tank easily, so keep your tank as full as you can. Don't wait until it's almost empty before you gas up again, especially if your car sits undriven for long periods — condensation and scale build up in an empty tank. Eventually, bits of corroded metal will work their way into fuel lines.

Chapter 4
Do-It-Yourself Repair

Making your own rust repairs — "doing it yourself" — saves money, and lots of it. Today even the simplest rust repairs run $75 and up in a body shop. And many body shops will refuse to repair rust at all unless it involves extensive body work.

Doing it yourself also saves time. Moreover, you can get started as soon as you spot rust. People who farm the work out often have to wait a while before a shop can take their car. In fact, some of the better body or restoration shops have waiting lists that are months long. In the meantime, that rust keeps spreading.

When *you* make a repair, you can count on quality. You know when you do a thorough job. But when you depend on other people, you have to take their word for it. And with rust, a job poorly done can lead to bigger problems. For instance, if you hire a shop to repair and repaint a rusty fender, and the mechanics fail to remove the rust completely before painting, the rust will break out again. And you will have to strip the metal, eliminate the rust, and then repaint again.

Finally, as you already know if you are accustomed to making mechanical repairs yourself, taking care of your own car keeps you up-to-date on its overall condition. You get in the habit of inspecting it regularly. And you begin to spot potential problems faster than ever. You also develop a greater appreciation for the complex invention your car is, with its thousands of mechanical and structural parts.

Nevertheless, there are circumstances under which it would not be wise for you to attempt your own repairs. For instance, if you have little or no practical experience in body work and the necessary repair is complex or extensive, you should farm it out. Or if it requires the use of tools or machinery completely unfamiliar to you, farm it out. Hire someone else ... this time. But practice and refine your skills in the meantime.

There are also preliminary repairs you can make yourself before handing your car over to a more qualified technician. These are described in more detail later in this chapter. If you hire someone else to do structural welding, you can do the body work to finish the repair. Or you can do some parts of the repair and farm out the rest.

Before you attempt any repairs, however, even the most minor repairs, make sure your working environment is safe and properly equipped. See the chart on page 53, which outlines some primary safety measures.

USING ABRASIVES

Almost any repair — even the simplest touch-up — will require the use of an abrasive. "Abrasive" is the general name for the gritty materials used to smooth metal, wood and other surfaces. In auto body repair, this process usually falls under the general heading of "sanding," but it can be broken down even further into "grinding" and "finishing."

The coarseness of abrasive materials is measured in terms of "grit." Number 16 grit is the coarsest; the finest is No. 600.

Grinding precedes finishing. *Grinding* removes paint and flaking metal from a panel that is to be repaired. Afterward, the panel is *finished* to remove any gouges and scratches remaining from the grinding process and to smooth it to a high-gloss surface ready for paint.

SAFETY RULES

- Wear safety goggles whenever you use machinery or compressed air — even a drill or sander.

- Never direct a stream of compressed air at another person or at skin or clothing.

- Secure long hair and loose clothing when you work with machinery or welding equipment.

- Keep small children out of your work area.

- Wear thick-soled work shoes to protect your feet from sharp objects, and long pants to protect your legs from jagged edges and sparks.

- Keep paint thinners, rust removers, paint and other chemicals away from your skin, especially your face and eyes, and wash with soap after every use.

- Always keep water and a dry chemical or CO_2 fire extinguisher handy in case of fire. Dry chemical extinguishers are best. Foam ones are too messy. Match extinguishers to the type of fire possible.

- When you work under a car, position it securely on jack stands; never rely solely on a jack.

- Keep all electrical cords in good condition.

- Store chemicals in original containers away from welding equipment and flames, even pilot lights. Don't keep lots of flammable liquids on hand.

- Keep your work area clean and free of obstructions.

- Never grind or weld near a battery. Either remove it from the area completely or disconnect the cables to help prevent shorts and explosions.

Grinding discs come in grits ranging from No. 16 to No. 80, and are either open- or closed-coat. The abrasive particles on open-coat discs are spaced further apart than they are on closed-coat discs. The wide spacing prevents the disc from clogging with paint too quickly. However, open-coat discs may leave deeper scratches on the surface.

Production paper is also used to remove paint and body filler. It ranges in grit from No. 36 to No. 100 and is generally of the open-coat type.

Finishing paper is available in a wet/dry waterproof version or as a dry paper. The dry paper is nonclogging, but it produces much more dust than the wet paper. Both come in grits ranging from No. 80 to No. 600, but most body men use paper in the 220-500 range.

When I mention "sanding" in this chapter, I mean the use of production or finishing paper, not grinding.

SMALL-SCALE RUST REPAIRS

If you take good care of your car, most of the repairs you will have to perform — if any — will be short and simple. Small-scale rust is probably all you will ever have to face. Chances are your first job will be repairing a tiny nick or chip in the paint surface. Other frequent, low-intensity repairs include fixing a rusted battery tray or corroded headlight buckets. Whenever you wash your car, check for such damage and repair it promptly, following the guidelines below.

Touch-ups. Buy a bottle or spray can — but preferably a bottle — of automobile touch-up paint that matches the color of your car. If your car is less than five years old, you can generally buy the paint by color code. You will find the code for your car on a plate mounted inside the passenger or driver door or under the hood.

If your car is older, you may not find such a code. If you don't, you will have to match the color as best you can. (If your car has a base-coat/clear-coat finish, spray on a clear coat after touching up the color. If your car has a metallic finish, skip this section. Consult your owner's manual instead.)

Some body paint *supply* houses will mix paint to match your car if it is older or of an unusual or discontinued color. Charges for mixing a pint or a quart average $9 to $15, depending on the type of paint — lacquer or enamel. Simply take the supply house a small, painted piece of your car — the door to the fuel tank, for instance. If your car is several years old and has not been waxed regularly, apply a light polishing compound to the surface beforehand. This will bring up the color and make it easier to match.

However, even if you carefully match the original color when you buy touch-up paint, you will probably not attain a perfect match, particularly on touch-ups an inch or more in diameter. That's because the paint on your car is constantly aging and maturing, and getting exposed to light and rain and other elements. A seamless match is not always possible. Besides, your primary goal in touching-up nicks and chips is to restore the protective barrier and prevent rust.

I've been lucky. I touched-up a nonmetallic finish three years after it was painted. I used extra paint I had mixed when the car was painted. I also used an inexpensive air brush, the kind available at hobby shops. The touch-up blended into the surrounding finish perfectly.

If the area to be touched-up is small and no bare metal has been exposed, clean it thoroughly and apply a dab of paint with a small camel-hair brush. Or use the brush that accompanies most bottled touch-up paint. Air brushes also do an excellent job.

If you have recently used a wax containing silicones or coated your car with polymers, remove them completely before attempting to paint or the paint will not adhere to the surface. Be thorough because waxes and polymers can be tough to eliminate. You can find special wax removers at some body paint supply houses. These will remove silicones and maybe polymers as well. Ask the salesperson if he or she has any cleaners developed especially for polymer finishes.

If you buy spray touch-up paint, spray a small amount in a cup and let it thicken a few minutes. Then apply it with a small brush.

For larger areas, or if bare metal has been exposed, more work is necessary. Using a fine wire brush, remove all traces of rust. A drill with a wire-brush attachment speeds the job. Then sand the area thoroughly, using a medium-coarse sandpaper. Make sure you get the edges, too. Sand again with a medium-fine paper.

Clean the metal with a tack cloth, then apply two coats of primer. Let the car sit for at least one night so the primer can cure. Then apply just enough touch-up paint to coat the chipped area — no more. That way the contrast between the original paint and the touch-up will not be too glaring.

Later, you might want to apply rubbing compound *lightly* around the edges to help blend the paint. Then wax the entire car. But never apply an abrasive or wax immediately after painting. It's best to wait several weeks. Also avoid painting when the weather is very humid.

Bubbled paint. Nine times out of 10, a bubbled paint surface indicates the area is rusting from the inside out. But bubbles can also occur when an improperly prepared surface has been covered with paint. If there are only a few bubbles relatively close to one another, chances are the damage is not too extensive.

Grind the bubbles off and sand down to the metal. With a small metal-working hammer, tap the surface lightly to determine where rust has penetrated and where the metal is still sound. Cut off all metal that has rusted through. The panel will have to be patched with fiberglass, body filler or metal (see page 61). Be extra careful if you use fiberglass. Most fiberglass repair kits require you to mix the resin with a catalyst before application. This catalyst can cause blindness if it gets in your eyes. Wear goggles whenever you mix these chemicals.

If all the metal is still sound, clean the inner surface thoroughly, if possible (use a wire brush). Then apply rustproofing solution to the same surface. (This may not always be possible, especially on a rocker panel.) Examine the area after you have cleaned it. If there is pitting, the rust is likely to return.

Sand the rust spot on the outer surface with medium-coarse and then fine paper, as explained above. When you finish sanding, there should be no visible scratches or other imperfections. If there are, the paint will only magnify them.

Go over the repair with a tack cloth. Then mask off the surrounding area and apply two coats of primer. Let the car sit overnight or longer, and then paint it with high-quality auto paint — not spray paint or touch-up paint, unless the area is very small.

Rusty trunks. If the floor or sides of your trunk begin to rust, water is working its way in somehow — probably through worn spots in the weatherstripping or through cracks or holes in the metal itself. Find the source of the water and eliminate it before you repair the rust. If you don't, you are just wasting your time.

Next, grind off any loose or flaking rust and clean the repair area with a wire brush until you get down to the bare metal. You can use an electric drill with a wire-brush attachment to get into tight spots or to speed the job. If the area still has some paint coating it, grind it off with a No. 36 disc. Then go over it again with a medium-fine paper, until there are no scratches or imperfections left. Go over the area with a tack cloth.

Before applying the primer, mask off the surrounding surface to protect it from paint. Apply two thick coats of primer. Allow it to sink into any cracks and crevices. Leave the car overnight or longer, and then paint.

Because the inside of the trunk is not a highly visible area, you can use a can of spray paint to coat it if you wish. Follow the instructions on the label. Always use the proper mask when working with spray paints.

The battery. The area around your car's battery is especially susceptible to rust. That's because of the acids that continuously spill out of the battery and corrode the surrounding metal. The high temperatures the engine generates and the area's vulnerability to moisture also contribute to rust. If your battery tray is in very bad shape, it will be easier to replace than to repair. Just remove it and install a new one. (If your tray is welded to the engine box, have someone cut it off and weld a new one on.)

Remove the battery terminals, negative cable first. Mix baking soda and water to form a paste and clean the terminals with it using a small brush. Scrape any corroded metal off the terminals and battery posts using a small knife or similar tool. If terminals are the screw-clamp type, disassemble them completely and clean the cable ends with the paste and a brush. You can also gently scrape the cable ends with the knife.

Check the cables at the ground end and clean them, if necessary. Then reconnect them — the ground end, that is. (When you are completely finished repairing the battery tray and are ready to reconnect the cables at the battery end, always connect the positive cable first.)

Remove the battery from the tray; wash it off and set it aside. If the tray is only lightly damaged from corrosion, here's what to do.

1. Grind off any loose or flaking rust and clean the area (down to bare metal) with a wire brush. Don't forget to look under the battery tray.

2. Sand the area with medium-coarse and then fine-grade sandpaper.

3. Degrease the repair area and the metal immediately adjacent to it.

4. Mask off any surfaces you want protected and apply two coats of metal primer. Let the primer cure overnight. The only job left is painting.

The engine compartment is subjected to high temperatures and corrosive elements. So you should paint the battery tray and the surrounding inner fender area with the same type of paint that coats the rest of the engine box. Or look for paint specially formulated for battery trays. (See Appendix 1 for details about tools and products that will make your work easier.)

Don't use touch-up spray paint. And never apply paint with a brush. For this job, a miniature spray gun works best (and can be rented).

Headlights. Headlights sustain a lot of damage over the course of the year. Being up front, they are first to get spattered with snow and dirty water. Research shows that the metal around the headlights and front bumper corrodes faster than most other parts of a car. So one of your first rust repairs may involve your headlight buckets — the headlight housings named for their bucket shape.

Remove the headlight assembly and dismantle it. First remove the retaining rings (sometimes called bezels), usually attached with Phillips-head screws. Then disconnect the tension springs that hold each headlight in place for aiming. If the screws are corroded or frozen in place, spray them with penetrating fluid.

Scrape off loose or flaking rust and clean the bucket with a wire brush — inside and out. You can stabilize the bucket in a vise to make it easier to work on. If it needs sanding, use a medium-coarse and then fine-grade paper. Then apply a good degreaser. Professional glass beading or sandblasting might be an alternative.

Apply two coats of metal primer and leave the buckets overnight. Then finish them off with a coat of spray paint that contains polyurethane. When the buckets are completely dry, put the assembly back together. Clean the lamp plug with a wire brush if it is corroded at all.

Check the gasket. If it is worn, repair it with silicone sealant or buy a new one, but make sure it forms a perfect seal. Replace any rusted screws with grade-8 hex-head or socket-head screws (in case the heads get stripped in the future).

UNDERSTANDING BODY FILLERS

Before you patch any holes or straighten any dents, you should understand body fillers. Most fillers available in auto parts stores are formulated so they can be used by people with wide-ranging levels of ability. They are formulated so even a beginner can mix and apply them with little or no trouble.

Most — if not all — are plastic fillers. More expensive and harder to find are aluminum-based fillers, which some reputable body shops use. A pint of plastic filler costs about $5. The same amount of aluminum-based filler can cost up to $20. Nowadays, plastic fillers are suitable for most home repairs, and are durable and easy to handle — as long as the metal is prepared correctly before they are applied, and as long as the application itself follows the directions listed on the product label.

Putties are different from fillers. They are used to fill pin-holes or fine scratches after the metal has been coated with a primer-surfacer and sanded.

BODY FILLERS AND PATCHES

Fillers. These substances fill irregularities in the body and can be sanded smooth. They also make the repair water-resistant. Fillers get combined with a cream hardener, and most are designed to make mixing as uncomplicated as possible. You'll have to really scrimp or really overdo it with hardener before any problems arise. After the filler and hardener are blended thoroughly, the substance is applied directly to the metal surface of your car and left to dry. Then it's sanded, reapplied if necessary, primed and painted.

Epoxy clay comes in a package containing a stick of epoxy and a stick of hardener, to be mixed in equal amounts and blended thoroughly. It can be molded to cover very small holes or minor breaks in the surface. Once it's dry, it gets sanded, primed and painted.

Fiberglass patches. Presized, prepackaged, adhesive-backed fiberglass patches make a good repair base. Simply cut the patch to size and stick it to the repair surface. Then cover it with filler, and sand it down to match the contours of the car. Follow with primer and then paint.

Aluminum repair tape. Once a repair area is stripped of rust and paint, it can be coated, in strips, with aluminum repair tape. The tape is then sanded and covered with filler.

Fiberglass repair kits. These kits are designed for the repair of small holes and cracks in fiberglass panels. Each kit contains resin, hardener, fiberglass cloth (or strands), and instructions. Be careful when you mix the resin, which is highly toxic; stick to small batches.

MAKING CORROSION WORK FOR YOU

Galvanic corrosion occurs when two different metals adjacent to one another are exposed to oxygen in a moist environment. The metal that is more susceptible to rust will give way first — "sacrifice" itself in place of the other metal. The auto manufacturers use this to their advantage when they galvanize sheet metal. They coat it with zinc, which is even more susceptible to corrosion than steel is. Then, when the paint surface is broken and corrosive elements get in, it is the zinc coating — and not the steel itself — that gets eaten away.

It's always better to use metals that have similar rust potentials. That's because a highly vulnerable metal and a fairly stable metal will set off an electrochemical reaction when they are placed next to each other. That reaction actually accelerates the corrosion process: the vulnerable metal deteriorates even faster than it would have otherwise.

Keep this principle in mind whenever you make repairs. Remember that something as minor as a screw can cause big problems if it happens to be made of the wrong metal. If you use brass bolts or fasteners on steel sheet, the steel will corrode at a much faster rate than it would normally. And the brass bolts will remain intact. (See the chart on p. 63 ranking different metals by their susceptibility to corrosion.)

REPAIRING ALUMINUM

Because aluminum panels are cheaper and lighter in weight (and therefore more energy efficient) they have become more commonplace. However, since they are softer than steel panels, you should be extra careful whenever you grind, hammer or sand them. When grinding, start with a No. 36 open-coat disc. Be careful not to gouge or cut through the metal. Try not to overheat the metal, either. Unlike steel sheet, aluminum does not change color before it melts; it just falls away. So when you work with

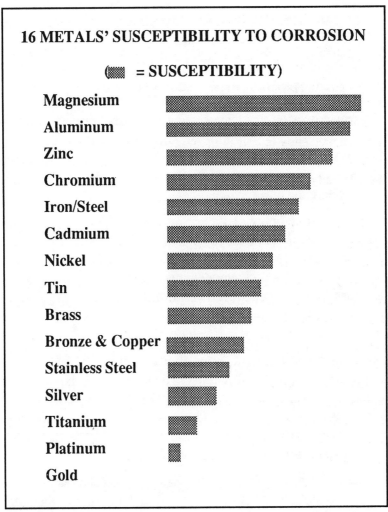

16 METALS' SUSCEPTIBILITY TO CORROSION

(▓ = SUSCEPTIBILITY)

Magnesium
Aluminum
Zinc
Chromium
Iron/Steel
Cadmium
Nickel
Tin
Brass
Bronze & Copper
Stainless Steel
Silver
Titanium
Platinum
Gold

an electric grinder, pause every couple of minutes to let the metal cool.

When you sand, never use any grit coarser than No. 80.

Body fillers will work on aluminum as well as steel sheet if the surface is properly prepared prior to application. Make sure it is clean and free of grease and oxidation. When sanding body filler on aluminum, use the coarser grits on the filled areas only. Stick to No. 100 grit for feather-edging.

Hammers, picks and other body-working tools can be used on aluminum. Just remember that it is a soft metal; it can't resist as wide a range of stress as steel can.

REPAIRING MEDIUM-SCALE CORROSION

Any corrosion that leaves a hole — no matter how small — can be considered medium- or large-scale corrosion. That's because the hole cannot be repaired by just sanding and painting. Extra steps are added: patching the hole and covering the patch, or replacing the entire panel.

Small rust holes (2 inches or smaller). Most repair jobs begin the same way: the area must be washed. Then remove any wax or silicone finishes within 12-18 inches of the rust hole and apply a good tricloroethane degreaser. Next, cut or break away any rust-weakened metal you find around the hole until you have only healthy, solid metal. Don't be surprised if the hole is twice its original size when you finish doing this.

Consider a rust hole an opportunity to inspect adjacent areas. Auto parts stores have slim mirrors on telescoping handles (like car antennae) to help you see into tight spots. You can also buy inexpensive lipstick mirrors. I have seen some fancy inspection lights. But you can easily make one out of a 12-volt side marker light bulb to look inside rocker panels.

Clean as much of the inside surface as you can reach, using a wire brush. Then apply a rustproofing solution. On the outer surface, grind away all paint 2 inches around the hole using No. 36 paper, but don't overheat the metal. With pliers, bend in the edges of the hole, then depress the entire area 1/16 to 1/8 inch. This allows enough room for you to install a patch and apply body filler.

Cut a patch to fit the hole. Use scrap sheet metal. Just make sure the patch is in good condition, clean, and free of any grease or paint. The patch should cover the hole entirely, extending from good solid metal on one side to good solid metal on the other. (The chart on p. 61 lists other materials that can be used to repair small holes. But the procedure I describe below involves the use of steel sheet and body filler.)

Next, mix the filler and hardener on a hard plastic or metal surface. Don't mix them on cardboard or any other porous surface that could absorb hardener. If you do, your mixture will be out of balance. Knead the filler and hardener together gently, trying to avoid air bubbles, until the mixture is uniform in color. Then use a little dab to attach your patch to the rear of the rust hole.

Next, using a plastic applicator, apply filler to the depressed area firmly (on the outer surface), expelling any air that gets trapped. Overfill the area slightly. When the filler is no longer rubbery — but is not yet completely dry — shape it with a "cheese-grater" file (see the diagram on p. 66). This will make sanding easier, and remove the resinous film that often clogs sandpaper.

When the filler is completely dry, grind the area with a No. 36 open-coat disc. (The filler label should specify how long it will take to dry; it is usually about half an hour.) Some fillers will feel a little sticky even when they are dry because of the resin they contain. Wear goggles and a safety mask because plastic fillers kick up a lot of dust when they are sanded. Then sand with No. 180 wet/dry paper until all nicks and scratches are gone.

Sand with the contour of the car, not against it. If you discover irregularities in the filler (like low spots that were not covered), apply another very thin coating of filler. Then sand with the wet/dry paper. Finally, feather-edge the repair area. To feather-edge, use a very fine paper to sand around the edges until the repair blends easily and invisibly into the rest of the car.

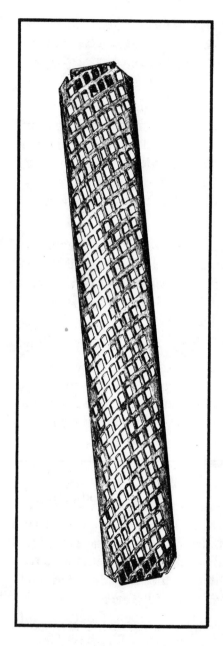

"cheese-grater" file

Once the area has been rubbed with a tack cloth, it is ready for priming. Mask off the surrounding metal. Then apply two thick coats of primer and allow the car to sit overnight or longer. All that remains is painting.

Repairing minor dents. Rust often develops when the outer surface is bent or dented in an accident. In such cases, the dented metal and the area immediately adjacent to it are stressed beyond their limits, making them especially vulnerable to corrosion. The protective paint coating also gets cracked or chipped. So when moisture and harmful elements like salt or acids come in contact with the dent, the metal immediately begins to deteriorate. If the area is left unprotected, the corrosion will spread beneath the painted surface to other parts of the car.

Few people realize they are capable of repairing minor dents themselves. They think only body shop mechanics have the expertise to do the work. Of course, for major body damage — mangled fenders, twisted panels and the like — a body shop is probably your best bet. But never neglect minor dents because you are reluctant to pay high labor costs for what seems to be an unimportant repair. Consider doing the work yourself.

Before you do anything, however, examine the dent and try to determine how the damage occurred. Go over the metal surrounding the dent. Often the force that dents the metal also shifts the surrounding area. See if you can pinpoint the direction of the force that caused the damage. To fix it, you will have to apply force in the opposite direction.

Was the panel struck at a perfect 90-degree angle? From above or below? Was it struck more than once? These are some of the questions you should answer before plotting your repair strategy.

There are several inexpensive tools on the market that you can use to straighten dents. For less than $10, you can buy a set of pull rods — rods with a handle on one end that are used to pull out dents. To straighten a dent with pull rods, you must first drill a series of holes (approximately 1/8 inch in size) along the bottom of the dent. Space them 1/4 to 1 inch apart (making them closest in the deepest part of the dent).

When you use pull rods, you usually begin at the shallowest part of the dent and work your way to the deepest. However, some dents will respond more quickly to the opposite approach. That is, some dented metal may pop back to its original shape if pressure is applied to its deepest point. You will have to judge for yourself which strategy will be most effective. Normally, however, you begin by inserting two or more pull rods in the holes you drilled at the shallowest part of the dent.

Pull outward — in a direction opposite the force that caused the dent, if possible. But pull gently. Don't try to straighten the whole dent in one go. It may take several rounds of pulling before the metal resumes its original position.

As you pull, you may also want to tap the metal along the dent with a small hammer to lower the creases. But tap lightly! Start at the shallowest part of the dent and work your way to the deepest end.

Another inexpensive tool, the slide hammer, works on the same principle as pull rods. This tool looks more like a screwdriver than a hammer. To use it, drill a series of holes in the bottom of the dent, exactly as you would for pull rods. Then insert the end of the slide hammer, which houses a screw assembly, into the hole at the shallowest point of the dent. (Again there are exceptions to this rule; on some dents you would pull the deepest point first.) When you pull the metal outward, a weight on the shaft of the hammer slides backward in the direction of the pull. This increases the corrective force. Work your way to the deepest end of the dent.

One company offers a "dent removal tool" for about $30 (see Appendix 1). It consists of a drilling tip and backing plate that attach to a 3/8-inch power drill. First, a hole is drilled in the damaged panel. Then the drilling tip is inserted. As the tip rotates, it pulls dents outward, against the backing plate. When the metal returns to its original contour, the tip automatically releases its grip.

After you pull out a dent with a slide hammer, pull rods or the dent removal tool, seal the holes you drilled. Use a solder gun or body filler. You may also have to add some filler to the repair in general (see below).

Many body men remove minor dents and straighten damaged panels with a hammer and dolly. A dolly is a small, curved piece of smooth metal that is hand-held on the underside of the panel being hammered. In one method, the dolly is placed directly beneath the damaged area, and the hammer strikes lightly from the opposite side.

The hammer blow itself lowers creases and high spots on the outer surface. It also causes the dolly to rebound slightly on the inner surface, helping to raise any low spots. This method — the on-dolly method — is usually used for pin-size indentations and very minor surface damage.

If the panel has a series of high and low spots — like ripples or waves — the dolly is placed behind a low spot and the closest high spot is pounded from the other side (see the diagram on page 71). This is called the "off-dolly" method.

Regardless of the method you use, you should always hammer the outer surface, never the inner one. And the hammer should strike the metal about 60 times a minute, with light, glancing blows.

There are many kinds of hammers and dollies available to the do-it-yourselfer (see Appendix 1). Some are general-purpose dollies. Others are more specialized. Most body-repair mail-order catalogues carry a wide assortment. But I found an adequate selection at a local auto parts store that caters to body mechanics.

The most important point to remember about using a dolly is that it should match the contour of the metal to be repaired. If you have a rounded dent, don't use a dolly with a straight, square face. Choose a dolly rounded to the same degree. Then, when you pound on the metal from the other side, you won't create more creases or gouges.

Likewise, make sure the hammer you use fits the job at hand, and that the face of the hammer is free from nicks and gouges. If you don't, you will simply transmit those gouges onto the car when you pound on the metal.

If you are just starting out, I recommend you buy a combination flat-face/pick body hammer and a good, general-purpose dolly. You can add to your collection as you go along.

It isn't always possible to use a hammer and dolly, however, since the undersides of many car panels are not accessible. Take the car door, for example, which consists of an inner and an outer panel. A "spoon" might be more practical (than the dolly) for maneuvering in tight spots in the door. A spoon is a long, flat, curved piece of metal, usually with a long "handle" that allows it to be held in place from below or above the panel.

Before you try to remove a dent using a hammer and dolly, grind all paint from the repair area, and all paint 2 inches around the repair. Use a wire brush to get into crevices and corners. Then remove any undercoating — that is, sound-deadening material, rustproofing or sealant.

It may be necessary to burn it off by heating the metal with a torch or other flame. Or use a chemical remover. (But never use a flame after a chemical has been applied.) A drill with a wire-brush attachment may work on lightly coated areas. You can even scrape the coating off manually if you have to. If you don't remove it all, you won't be able to control the effects of your hammering; the undercoating may distort the shape of the panel when the panel is pounded. (Also remove any body filler or putty that may have been applied to the panel.)

When the dolly is placed on the underside of the metal, it is out of sight. So you may have to practice placing it behind the damaged area, or behind the low spot you want to raise. (It isn't as easy as it seems to match it up with the hammer blows.)

After you pull out a dent, use a hammer and dolly to smooth the area as much as possible. If you do the job well, you should end up with a panel smooth enough to sand, prime and paint without body filler.

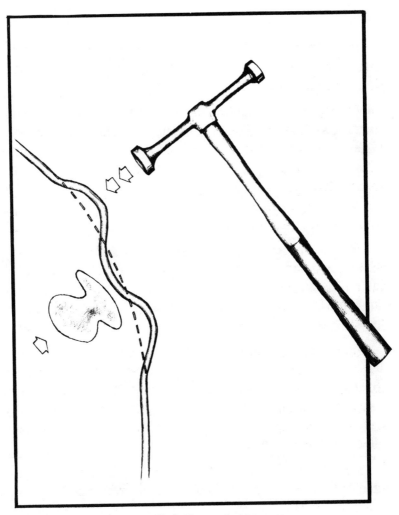

off-dolly method

Usually, however, you will have to apply at least some filler. But remove as much of the dent as you can first. Get the panel as close to its original shape as possible. Then clean the area with a good tricloroethane degreaser and go over it with a tack cloth.

Mix the filler and hardener on a hard plastic or metal surface (not cardboard; see the explanation above). Then apply it to any low spots on the panel, pressing firmly to expel any air that gets trapped. Slightly over-fill the low spots.

When the filler is no longer rubbery — but is not yet completely dry — shape it with a "cheese-grater" file (see the diagram on p. 66). This makes sanding easier and removes some of the film that clogs sandpaper.

When the filler is completely dry, grind the area with a No. 36 open-coat disc (the resin will clog regular paper quickly). Wear goggles and a safety mask because plastic fillers kick up a lot of dust when they are sanded. Then hand-sand with No. 180 wet/dry paper until all nicks and scratches are gone. A narrow, two-hand sanding block — often sold under the brand name "Bondo board" — is a useful tool to have on hand for sanding filler (one-hand blocks are also available for tighter spots).

Sand with the contour of the car, not against it. If you discover irregularities in the filler (like low spots that were not covered), apply another, very thin coating of filler. If you discover high spots you missed during the hammer-and-dolly stage, stop and hammer them down immediately. Then reapply filler where it's needed and let it dry.

Sand again by hand, using wet/dry paper. Finally, feather-edge the repair area. Now it's ready to be primed and painted.

Patching a panel. Small rust holes grow into afflicted panels in no time if they aren't repaired. Then the bad metal must be cut away and patched. However, for these bigger jobs — in contrast to small rust holes — the patch must usually be welded in place.

"Dimple tools," available from at least one major mail-order house, allow you to install a patch without welding — or at least without having to depend primarily on welding to hold the patch in place (see Appendix 1). Or you can patch a panel yourself using screws and a dimple tool, and then have a professional welder finish the repair. This will save you money.

One dimple tool looks like a C-clamp. Another looks like a pair of locking pliers. When it is tightened against the metal, it creates a dimple, a small indented circle. A hole is drilled in the center of the circle, and a screw is inserted to attach the patch to the original metal.

Before you install a patch using a dimple tool, determine whether the repair involves a structural area. A patched panel will be stronger than a rust-weakened one, but will it be strong enough to lift the car if it involves a jack-up point? Will it be strong enough to protect you in a collision? If you have any doubts, consult a professional body mechanic. But if it involves simple nonstructural rust, follow the steps below.

1. Clean the surface of the metal and remove any wax and silicones.

2. Cut out any metal that has been weakened by rust. When you are through, the hole may be twice its original size — don't worry; that's only normal. Inspect the surrounding area (12 inches) for signs of rust.

3. Clean the inner surface of the metal with a wire brush, if possible, and apply a rustproofing or rust-treating solution. However, if welding is to follow your repair, you can expect rustproofing and other chemical treatments to burn off in the immediate area. So skip this step.

4. Grind off all paint 2 inches around the hole on the outer surface using No. 36 paper. Pause every few minutes so the metal doesn't overheat.

5. Make a patch for the repair area out of sheet metal or aluminum or zinc. The patch should overlap the original metal by about 1/2 inch. Be sure the patch is of the same gauge as the rest of the metal on the panel (or a similar gauge).

6. Make dimples around the edge of the hole — but not in the patch — using the tool. Space the dimples 2 inches apart and 1 inch from the edge of the hole. When you finish, center-punch the dimples and drill 3/32-inch holes into them. The holes must be placed exactly in the middle of the dimple.

7. Attach the patch to the metal temporarily, from behind. You can attach it by inserting two screws diagonally across from one another in two of the dimples. Once the patch is attached securely, so it doesn't wiggle at all, insert a pencil or marker in each dimple and mark the patch behind. Then remove the patch and drill a 3/32-inch hole at every mark.

8. Reattach the patch by inserting No. 6 panhead Type A sheet metal screws in the holes. (If you choose to use pop rivets instead of screws, drill holes the appropriate size.) As you tighten the screws, the patch will be drawn in against the metal. Make sure the heads of the screws are below surface-level when you are through. Don't use brass or stainless steel screws or they could cause the surrounding metal to rust.

9. Apply body filler to the low spots (every dimple) and let it harden.

10. Sand the surface. Now it's ready to be primed and painted.

Rust that extends across a panel to the edge of the panel should usually be patched using the welding process. But the dimple method is fine for many other jobs, provided you rustproof or rust-treat the inner surface before installing the patch. I outline the steps involved in welding a patch in place later in this chapter. First, however, I describe some welding terms and processes.

SOME WELDING BASICS

If two melted pieces of metal are pushed together, they will form one continuous piece of metal when they cool. This is a basic description of the welding process: enough heat is applied to metal to melt two or more pieces together, and a stream of "filler" metal is melted at the same time to ensure a solid joint.

Most people who weld, weld for a living. But more and more do-it-yourselfers are acquiring the skill because it simplifies so many jobs. If you know how to weld, you can design and fabricate your own mechanical parts and make a host of different repairs. In fact, the difference between a good body mechanic and a mediocre body mechanic is usually the ability to weld *well*. That ability is especially important if you own or collect vintage cars. It enables you to make substitute parts you might not be able to find otherwise.

There are several different types of welding. Among them are oxy-acetylene welding, brazing and electric welding. For rust repairs on sheet metal or aluminum, electric welding is your best bet.

Electric welding includes arc welding, and MIG (metallic inert gas) and TIG (tungsten inert gas) welding. Arc welding is usually reserved for jobs on heavy steel, like automobile frames or trailer hitches. That's because most arc welders cannot be adjusted to amperages low enough for sheet metal and aluminum.

Because of this, most professionals use MIG welders for body work. MIG welders cost $300 or more — and are highly recommended for auto repair. In fact, no other type of welder can surpass the MIG in body repair.

Now, however, at least one major mail-order source of do-it-yourself body tools and equipment offers a home *arc* welder. It costs less than $200 (including a face shield) and can be adjusted to amperages of 50 and under. In addition, the same source offers spot and stitch welding attachments (see Appendix 1 for brand names and price ranges). If $200 sounds like a lot to spend on a welder, just remember that even one small rust repair requiring welding in a body shop can cost that much — or more!

This arc welder, which runs on regular house current, consists of a metal control box (the welding machine) with one ground cord and one power cord. The ground cord gets clamped to the metal to be welded, as close to the repair area as possible. The clamp at the end of the power cord holds the electrode. An electrode is a piece of coated metal wire (the "filler" metal mentioned above) that ranges in diameter from 1/16 to 5/16 inches and in length from 9 to 18 inches. (The electrode is sometimes called the welding rod.)

During the welding process, current travels along the power cable to the electrode. When the electrode is held near the base metal (the metal to be welded), an electric "arc" — so called for its shape — is formed. As the arc is moved along the area to be welded, the metals to be joined are melted and fused together, along with the electrode. When they cool, they constitute one continuous piece of metal.

More powerful welders may seem easier to use. But they often just disguise sloppy welding techniques. Because they generate more heat, they can cut through all kinds of metal — even metal with remnants of paint and rust. A low-amperage welder cannot do this. So you will have to perform all the steps involved in the welding process thoroughly, from cutting the patch to sanding and grinding it clean.

Low-amperage welders are not inferior to their more powerful counterparts. A regular electric arc welder would probably burn right through the thin metal used on cars.

As metal melts, it weakens and becomes more volatile. So it runs the risk of oxidizing — which is what occurs when it mixes with oxygen. Because of this, electrodes now come with a flux coating that produces a gas shield when it is burned. This shield protects the molten metal from impurities in the air. The

WELDING SAFETY

- Always wear a helmet with the proper lens when welding with an electric arc, and chipping goggles when removing slag from a weld.

- Never weld near paints, thinners or other flammable liquids.

- Never weld near a car battery. Remove it from the area, or disconnect the cables to prevent shorts and explosions. Be equally cautious about electrical wires and fuel tanks. Close fuel filler pipes tightly, and cover them with damp rags.

- Make sure your work area is properly ventilated. Metals coated with paint or chemicals give off noxious fumes during welding, especially galvanized metal.

- Keep your work area clean and free of clutter. Store combustible materials elsewhere.

- Never weld on a wet floor.

- Never position yourself directly under the welding area. Instead, stand or lie to one side to avoid sparks and molten metal.

- Wear heavy, long-sleeved shirts, long pants, leather gloves, and work shoes free of oil, grease and other flammable materials. Avoid woolly sweaters, short-sleeved shirts, and open pockets and cuffs, which can trap sparks and smoldering metal. Also avoid synthetic fabrics, which can melt against your skin.

- Keep water and a dry chemical or CO_2 fire extinguisher handy in case of fire.

flux also leaves a layer of "slag" over the weld, which protects the joint as the metal cools. Afterward, the slag gets chipped away.

Electrodes are labeled by a series of numbers. These numbers correspond, among other things, to the tensile strength of the metal to be welded (how many pounds of pressure it can withstand per square inch). They also indicate the welding position to be used.

The diameter of an electrode should not normally exceed the thickness of the metal to be welded. But if you can't find an electrode with a diameter small enough for the job at hand, you can compensate by adjusting the amperage or the speed at which you move the arc along the joint. This is explained in more detail below.

Many of the fundamentals of arc welding I describe also pertain to oxy-acetylene and MIG/TIG welding. I also recommend a book entitled *Auto Body Repair and Refinishing* by Robert P. Schmidt (published by Reston Publishing, a subsidiary of Prentice-Hall). It has excellent diagrams and detailed descriptions of the many components of these processes.

This section is intended to give you an overview of the welding process. I urge you to take a course on welding and to use all recommended safety gear.

Getting started. Go over the safety tips listed on pages 77 and 79, making sure there are no hazards in your work area. Welding can be very dangerous unless you follow the proper procedures.

First, always wear a face shield. The ultraviolet and infrared light given off by the electric arc (and other types of welders) can severely damage the human eye, as can flying sparks and molten metal.

MORE WELDING SAFETY

- Carefully follow the instructions that accompany the welding machine, especially those about its electrical system.

- Remove any upholstery, carpeting, or similar materials that may catch fire, or cover them with damp rags.

- Be extremely careful. Fires can break out easily during welding. Check for them frequently.

Second, dress properly. Many senseless injuries and fires can be avoided if you wear the right kinds of clothes.

Third, have a plan of action to follow in case of fire, and the proper equipment (like a fire extinguisher) to contain it.

Always approach welding respectfully and cautiously. Never attempt something you feel uneasy about; instead, consult a professional for guidance. If you are mindful of the hazards involved and take the proper steps to avoid them, you should be fine.

After you carefully review the safety chart, gather several pieces of scrap metal. Get the metal as free from grease, paint, rust, primer and body filler as you can. These coatings can limit the effectiveness (not to mention the ease) of the welding job. Choose an electrode suited to the thickness of the scrap metal. (Most welding supply houses will provide you with a chart listing the different electrodes available and their diameters, tensile strengths, range of currents, etc.)

The amount of current required for a particular job depends on the diameter of the electrode. And, as I explained earlier, the diameter of the electrode depends on the thickness of the metal to be welded. So the thinner the metal, the lower the amperage.

Remember, each electrode is designed for use in a particular range of currents.

Connect the ground cable to the base metal, and shield your face and eyes with the proper equipment. Then turn on the machine. Your first challenge will be striking the arc, which can be difficult at first if you haven't practiced. In addition, there are several methods to choose from.

You can hold the electrode at a 20-25 degree angle (to the base metal) and scratch its tip against the metal surface. Use the same motion you would use to strike a match. As soon as a spark appears, lift the electrode until its tip is approximately 1/16 inch above the metal (assuming the electrode you are using has a diameter of 1/16 inch). Don't lift the electrode too quickly or you will lose the arc.

Another method calls for you to hold the electrode at a 90-degree angle to the base metal and tap it against the surface. When you get a spark, slowly lift the electrode until it's about 1/16 inch above the surface (if your electrode has a diameter of 1/16 inch).

Or you can lay the electrode on its side with the tip extending over the edge of the base metal. As you drag the tip onto the metal, sparks should appear. Raise the electrode into the proper position.

Finally, you can drag the electrode across the metal a few inches at a very shallow angle. This will generate a series of sparks. You should be able to translate one of them into an arc, using the process described above.

If you fail to generate any sparks using these methods, your welder may be set at too low an amperage. Other possible problems include an improper connection at the ground cable, and excessive paint, rust or grease on the metal to be welded.

Striking an arc is complicated by the fact that you can't see anything at all when you are wearing a protective face shield. You can't see anything until you strike an arc, that is. So you will have to position your electrode before pulling down the shield and turning on the power. And you will have to learn to strike an arc by "feel" instead of sight.

Another problem that often plagues beginners concerns the electrode, which often gets stuck to the base metal. When this happens — and it will probably happen at least once while you are practicing striking an arc — you should gently wiggle the electrode holder and try to unstick it that way. If you can't get it free, release the electrode from the holder. Then lift your face shield and pull the electrode off the metal with a glove or pliers.

Examine the electrode. If the flux coating is burned or darkened, or if the electrode is bent or otherwise damaged, discard it. It will only cause more problems.

Establishing the proper distance between the electrode and the metal surface — the arc length — is critical. This distance generally corresponds to the diameter of the electrode being used, so it will vary from job to job. If the electrode is held too close to the metal, the voltage will be too low to melt the base metal properly. In addition, the bead — the line of molten metal left along the joint — will be too high and uneven, and the electrode may stick to the metal.

On the other hand, if the electrode is held too far from the metal, too much voltage will be generated, the arc stream will wobble, the electrode will melt in large globules instead of a steady stream, excessive spatter will result, and the base metal will not fuse properly.

The rate at which the arc moves along the joint also affects the welding process. If it moves too quickly, you will be left with a narrow bead, and the metal may solidify prematurely. In addition, the metal will be insufficiently penetrated. To get a strong joint, you must make sure that the heat from the electrode penetrates both metal panels. It must penetrate them enough to bond them tightly together.

On the other hand, if you move the electrode too slowly along the joint, the bead will be too wide, and too much metal will be melted. Penetration will be excessive.

There are several different ways to weld a joint. Many professional welders use a circular or weaving motion as they move the electrode along the joint. In this way, they manage to disperse the heat over a wider area and avoid distortion or burn-through. Distortion is when the plates bend out of line during the cooling process.

As a beginner, however, your best bet is probably the "drag" method: pulling the electrode along the joint in a straight line. Once you become more proficient you can try other techniques — like pushing the electrode or using a weaving pattern as you go along. But make sure you don't just heat the upper surface of the metal when you use a weaving motion — penetration is important.

Also, be aware that the position of the metal you are welding also helps to determine what type of electrode you use, the amperage setting you select, and your rate of travel. For example, if the joint you are welding lies in a vertical plane, you will have to adjust these factors to compensate for gravity, which will pull the molten metal downward with more force than usual. And if you are welding on the underside of your car, you may have to use a higher amperage and a slower speed to penetrate the metal panels sufficiently. The key is practice. The more you weld, the greater your understanding of the different elements involved.

Practice scratching or tapping an arc several times before you turn on the power. And practice welding pieces of scrap metal before you move to the actual repair. Practice is everything!

It will probably be a while before you can coordinate all the different parts of the process. The important thing is to move the electrode along the metal at a steady — not irregular — speed and in a manner that enables you to observe the bead as it forms. And if you are using a low-amperage welder, you will have to look much closer than normal.

When you are ready to weld the actual repair area, tailor the patch or panel you are installing to fit it. And then find a way to hold the patch in place securely while you are welding, using C-clamps, vise grips, etc. Eliminate all gaps between the original metal and the patch or panel.

One company offers "panel holders" that are easy to install — and easy to remove once the job is done. All you have to do is seal the holes they leave when the repair is finished, using soldering or spot welding techniques (described below). See Appendix 1 for a more detailed description of the products available.

The bead. The bead should be slightly raised over the metal surface, with even, semicircular ripples. The width of the bead will vary, depending on the width of the electrode you use. Narrow beads with pointy ripples usually indicate excessive speed. Very wide beads that are raised more than usual indicate too slow a speed.

The arc length can also affect the quality of the bead, as I said before. Insufficient distance will result in a narrow bead and excessive distance in too wide a bead.

Practice running beads on scrap metal. You can draw lines across the metal using a pencil made of soap-stone, which gives off no fumes when it is burned. When the metal cools, remove the slag and examine the beads.

Types of joints. In welding, there are four basic types of joints: the butt joint, the lap joint, the fillet or "T" joint, and the outside-corner joint. In car repairs, only the butt and lap joints are normally used. I will focus on them, although the diagram on page 86 depicts the "T" joint as well.

Butt joints. When two pieces of metal are welded together in a straight line, or one continuous metal plane, they form a butt joint (see the diagram). To obtain a strong and durable butt joint on metal plates 1/8 inch thick or thinner, space them a distance equal to half their thickness. For example, space two 1/8-inch plates 1/16 inch apart.

When your repair requires that you weld plates together in a butt joint with a long, continuous bead, you can avoid metal distortion by welding in segments instead of in one continuous line. After tack-welding the ends of the plates — welding in just one spot at either end of the joint — begin welding 2 inches in and stop when you reach the outer edge. Then weld another segment, starting another 2 inches in and moving the arc toward the first weld. Continue this procedure across the joint. Or follow this procedure, but skip every other 2-inch segment along the joint. Then go back and weld the segments you skipped. Practice making butt joints until you can achieve the proper bead without distortion and excessive heat.

Butt joints cause problems on especially thin metal, so you should use more lap joints. However, if one of the joints you butt-weld bends out of shape as the metal cools, you can usually straighten it out with a hammer and dolly.

Lap joints. The lap joint is one of the easiest and sturdiest joints to weld in body work. It results when two pieces of metal that overlap each other are welded together (see diagram). One variation of the lap joint — the flanged lap joint — is especially recommended for auto body work. It leaves a less visible seam and results in greater penetration.

Often, you will only be able to weld the outer surface of the lap joint. However, the joint will be stronger if it is fused on the inside as well.

Clamp the pieces of metal together tightly before welding, or the upper panel will get excessively hot. Hold the electrode at a 45-degree angle to the bottom metal plate and drag it along the joint. If the metal on top begins to melt away, aim the electrode at the bottom plate.

Spot welds. Once you have welded a patch or panel in place, hide all the holes you drilled. Remove the panel holders you installed. Then strike an arc, hold the tip of the electrode over one of the holes, and push downward gently. The heat should melt the bottom panel slightly. If you then withdraw the electrode slowly, it will melt and fill the hole.

Testing joints. You can test your practice joints by clamping the metal to a vise. But wait until the metal has cooled completely. Clamp the metal along the joint, not at the joint. Then, using pliers or a hammer, bend the metal at the joint. Twist it back and forth until the metal breaks. When it does, it should break *beside* the welded joint — not *at* the joint.

You can also cut the joint cross-wise using a hack saw. Then examine the weld. In this way, you get a good idea of the degree of penetration you are achieving and the type of bead you are leaving.

REPLACING BODY PANELS

At one time or another, you may be faced with rust severe enough to warrant replacing most of a panel. This job usually involves cutting away one or more of the panel's outer edges. Rust this bad is usually found on older cars, often on cars that are being restored. Repairing the damage involves more than cutting a patch to fit a hole. It normally means fashioning a patch

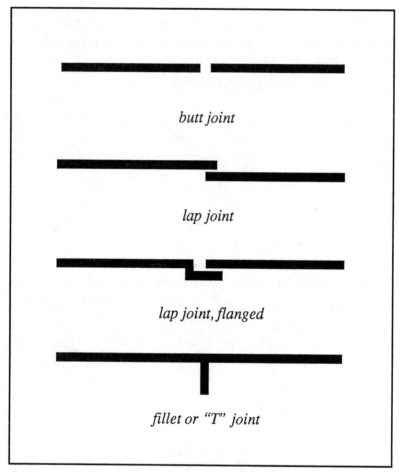

butt joint

lap joint

lap joint, flanged

fillet or "T" joint

that has a finished outer edge as well — like a wheel arch or bottom front fender.

Although this may seem like too big a job for a do-it-yourselfer, it can be simplified. For one thing, at least one company offers replacement panels by mail (see Appendix 1). You may also be able to find a replacement panel at a salvage yard. In addition, many shops that specialize in metal-working can fabricate a patch to fit your specifications. One day you may even be experienced enough at welding and metal-work to fabricate the part yourself.

To repair panels severely damaged by rust, test the metal around the damaged area. Cut away any that is not structurally sound. Use snips and a light metal-working hammer to do this. Second, find or fashion a replacement patch. This is the most important part of the job: making sure the patch fits like a glove. If it doesn't, you will probably overheat the patch or the original metal when you begin to weld. Metal distortion will result. And a patch cut to the wrong size will never match the original contour of the car.

If you order a panel by mail or get it from a salvage yard, you may have to cut away some of the metal on your car to accommodate it, or vice versa. The patch and the original metal should overlap each other only by about 1/2 inch.

One mail-order source specializing in auto body tools offers a panel "flanger." This is a device that crimps the metal around the edge of the repair so the patch nestles right down into it (the diagram on p. 86 shows a flanged joint; the panel flanger itself is described in Appendix 1). After such a patch is welded into place, the surface is smoother and there is less evidence of overlapping metal. But the edges should be stripped of paint and rust before they are flanged.

Once the car and the patch are cut to the right size, grind away any paint or flaking rust within 2 inches of the metal edge — on both the patch and the car. Get the metal as clean as possible or it will be difficult to weld. This is as important as fitting the panel snugly. Then apply a good degreaser and remove any wax or silicones as well.

The next step is positioning the patch. Hold it up to the repair area and clamp it in place with locking pliers or vice grips. Make sure there are no gaps or protruding metal, and that the patch will not move once you begin to work on it. Then drill 1/8-inch holes around the edge of the patch every 2 to 4 inches (the bigger the patch, the greater the space between the holes), approximately 1/2 inch from the edge. Next, insert panel holders (described in Appendix 1) in the holes. They will compress the two pieces of metal together tightly during welding. (Rivets are another option.)

Put on your protective face shield. Then, using your panel welder, tack the patch in place in four or five spots. This is in preparation for the final welding, which, in this case, involves a lap joint. Start at one end of the joint, weld a short segment, and stop. Do the same at the other end, gradually working your way to the middle. Remember to hold the electrode at the proper angle and to travel at a suitable speed. If you don't, your weld will be ineffective or, at the other extreme, excessive.

Once the weld has cooled, chip the slag away and grind the bead as smooth as possible. Apply body filler to any low spots or creases. Then sand, prime and paint. Follow the same procedures for sanding and priming that I described earlier in this chapter. Remember to feather-edge the repair when you sand. Apply two thick coats of primer, and allow the car to sit overnight or longer before painting.

Chapter 5
Older or "Classic" Cars

If you own an early-model car, you will probably encounter most of the problems described in this report. These include rust holes, minor or major dents, blisters beneath the paint surface, and chalky paint. You may have encountered them already! When these cars were built, the manufacturing process included little or no rust protection. In fact, it often *encouraged* corrosion. Seams were welded and left unprotected, chrome trim was bolted onto a contrasting steel body, drainage was haphazard at best, and so on.

If your older car is still in relatively good condition, you should implement an especially stringent maintenance program. This program should include frequent washes and waxing, regular underbody cleaning and inspection, and so on. On the other hand, if you buy (or have bought) an older car that needs considerable work, there are several basic steps you should follow to put it in order. Then the same stringent maintenance must be ensured.

First, however, if you plan a complete overhaul — or even if you don't — find a manufacturer's service manual for the make, model and year of the car. These are often advertised in catalogues of car parts and accessories, and in automotive magazines (especially those that focus on old or collectible cars). Such a manual will make all the difference in the world. In fact, it would be foolish to attempt a restoration without one.

Check with the sellers of auto-related books, especially the larger mail-order companies (see Appendix 1). They may have a restoration manual for your model and year of car. This can save you much time and money. Also look for a car club for your make and model. People who have restored a car may know the pitfalls and have hints to share.

The restoration process mirrors the manufacturing process for cars with a separate body and frame; everything is done in reverse, that's all. During manufacture, as the frame, or chassis, travels down the assembly line, the suspension units are added. Then the engine and the rest of the drive train are installed. Next, the painted body is dropped onto the frame. As it continues down the line, the car is equipped with its wiring and steering systems, and brake and fuel lines. Then the interior, window glass and gaskets are added. Finally, the bumpers and trim go on.

In restoration, you start with an assembled car and take it apart. A complete restoration usually involves the following steps:

- Gut the interior. This means removing the seats, trim, dashboard (if it's separate from the body itself), upholstery, headliner and rugs. Then remove wiring harnesses and glass, and the steering assembly and fuel and brake lines (where they are attached to the body). Catalogue all parts that are removed, including nuts and bolts. In other words, keep a careful record of where each part belongs in the final assembly.

- Remove the body from the chassis.

- Remove the drive train (it's easier once the body is off) and send it out for rebuilding.

- Steam-clean or power-wash the chassis and then degrease it. This process may take longer than you would expect, especially if there are years of caked-on mud and grease to remove. Take the chassis apart and sand it. Look for hidden rust.

- Send the chassis out to be chemically stripped, if necessary.

- Repair the chassis, if necessary, and check its alignment. Some experts think you should avoid a bent chassis at all costs. If your initial inspection failed to reveal a bent chassis, and you discover it at this stage of the restoration, you have many additional steps to cover; you should also be aware that it is very difficult — some say impossible — to straighten a bent chassis correctly. Consult a professional body mechanic.

- After the chassis has been repaired, you will want to prime and paint it. Powder-coating is another, longer-lasting option.

- Strip the body of paint, grease and rust. Then disassemble it completely. Or, if you can afford it, have the panels professionally stripped (see p. 98).

- Do all necessary body work.

- Drill any holes necessary to attach the trim and other accessories. This step will only be necessary if you are adding or changing exterior trim.

- Remove all loose caulking. (If you had your car professionally stripped, the caulking will already be gone.) Then replace it.

- Coat all body panels with primer, inside and out.

- When the primer is completely dry, block-sand the body. Then paint it, inside and out.

- Reinstall suspension units. Change all ball joints and bushings.

- Paint the engine and running gear and install them on the chassis.

- Reassemble the body. Replace all body mounts and rubber bushings between the body and chassis. Install the body on the chassis. Try on the new (or restored) dashboard (if it's separate from the body), pedals, steering column and exhaust system. Check the body for correct panel alignment.

- Install the brake lines and bleed them.

- Install the wiring, cooling system, etc.

- Install all glass and weatherstripping.

- Install the newly upholstered seats.

- Sand the painted body, if necessary, and apply rubbing compound. (That is, if you used acrylic lacquer to coat it and if the paint has cured two to four weeks.)

- Touch-up the paint, if necessary. Then wax the surface.

- Install all moldings, trim and bumpers, etc.

One step was omitted from this list, and for a reason. That step is sending chrome parts out for replating. Since restoration projects often take much longer to complete than expected, it's best to hold onto chrome parts until late in the process. If you have them replated too early, and have to store them while you finish the restoration, they may begin to corrode again.

As you can see, restoring a car is a massive undertaking. But once you finish, you can be sure the car is sound and sturdy. Then detailed maintenance checks can help you keep it that way.

Body work, sanding, priming and painting are basically the same for any car. But older cars are more sensitive than the average automobile. And that's why I have written this chapter — to fill you in on areas that should get extra attention.

STRATEGIC AREAS

Remember, older cars have a different structure. For one, they have a frame and a body (today's cars come in one unit, although some older foreign cars are also unibodies). On older cars, the frame is the foundation of the entire structure, and houses the power train and suspension as well. The body, consisting of the passenger compartment and all the metal panels that surround and enclose it, is mounted on the frame.

In addition, trim on older cars is usually metal — not plastic. And it is usually a metal other than steel. Trim and moldings are usually bolted to the body. Holes are drilled in the body panels to accommodate the bolts. A very old car (and some English and European models) may also have some parts made of wood, primarily in the passenger compartment.

I am reminding you of this to make a point. If you understand old-car construction, you can pretty well chart the course of any corrosion that develops. It will attack the "joints" first, where the frame and body come together, where body panels are welded together, where wood meets metal, and where one metal meets a metal of a different type.

It will also strike where the metal has been weakened by welding or by the drilling of holes, or in places where water runs and collects. So if you are restoring an old car, make sure these areas receive extra attention and protection. After you weld an area, coat it with a rust preventive. Then make sure water drains from it properly. If you must drill holes in a panel, coat them with a rust preventive and then with primer. If you have to replace any trim, use plastic or a metal that doesn't trigger corrosion (see the chart on p. 63).

In many instances, you may have to replace entire sections of the car because corrosion has eaten so much of them away. If so, cut out all rust-weakened metal. If you don't, the rust will just begin to spread again. And use a solid, rust-resistant replacement metal.

A car that has been neglected will probably have a corroded cooling system, fuel system, brake cylinders, and so on. All these must be repaired or replaced.

The frame. In restoring a car, start with the frame. Have it checked for proper alignment. If it is twisted or distorted even slightly, it can throw the whole restoration out of whack. For example, a frame that sways or sags can interfere with fender, hood and door alignment. It can even ruin wheel alignment. And, as I said before, it may be difficult — if not impossible — to correct.

Try opening and closing the doors a few times. Is there any play in their movement? Do they sag? Do they close easily and completely or do they scrape against the door frame? Also check the hinges for defects. If the hinges are OK, but the doors scrape or sag, the frame is probably twisted or deteriorating.

Wooden components. It is often difficult to judge the condition of wooden components on a classic car because they are hidden behind body panels or trim. But there are places where wood can be seen, though these places may vary from model to model. First, look beneath the dashboard. A wooden cross beam may be visible. Check under running boards for wooden chassis rails. When you find wood, try to mark its surface with a knife or tool. It should be impervious to such attempts. If it isn't, it may be dangerously weak.

Body trim and moldings. Trim and moldings on the car serve a variety of purposes. For one thing, they beautify the car. They also protect it. For example, body side moldings shield the metal from dents and dings. In addition, trim parts cover body seams or finish off openings. On most older cars, the trim is metal — aluminum, stainless steel, chrome. And both trim and moldings are attached to the body in several ways, usually involving the drilling of holes.

Before you begin any body repair or refinishing, remove the trim and molding. But be careful! Old metal trim may be especially brittle. That's why I told you to get a service manual. It will tell you, among a host of other things, how to attach and remove trim and moldings.

In many restorations, entire panels must be replaced. The original trim — or suitable replacements — must then be installed on the new panels. When you plan to transfer the trim from one panel to another, simply remove the old panel with all the trim still attached. You can even use the old panel as a pattern for the new one, so you will know where to drill holes. If you have to remove the trim, however, tape all the small parts together and label them clearly. That way you will know which parts belong together.

Remember that metal trim — especially if it is a metal other than steel — can trigger corrosion. Rivets used to attach trim to the body can do the same. So choose plastic trim or a safe metal (see the chart on p. 63). Moreover, when you reattach trim to a repaired panel, make sure water and dirt can escape around it. Otherwise, a rust hole will develop.

Finally, keep in mind that it is best to drill any necessary holes before you prime and paint the car. Otherwise, you could damage the paint coating.

The brakes. One of the first automotive systems to corrode is the brake system. That's because it is more of an appendage to the body — not an integral part it — especially on older cars. As a result, brake lines corrode when water, mud and slush splash against them. And the rest of the system corrodes because brake fluid absorbs moisture over time.

If the braking system on your older car is already corroded — and it probably is — you will have to replace the individual components. But if the system is still sound, you can protect it by flushing it completely every 18 months and refilling it with

new fluid — from an unopened can — or with silicone brake
fluid.

The cooling system should be handled in the same manner.
Just refill it with water and antifreeze.

Fiberglass bodies. Some cars — like the Chevy Corvettes
of the 1960s and 70s — have fiberglass bodies. They were
designed to end the threat of auto body rust. Unfortunately, the
rust-free bodies hide the same *rust-prone* steel frames underneath. These frames generally corrode from the inside out. And
they corrode quickly, usually because their unsuspecting owners
believe rust to be an irrelevant issue.

Before you buy a car with a fiberglass body, check the
integrity of the frame underneath (see Chapter 8 for some tips
on inspecting such cars). If it has been coated with paint or
another substance, rust may have already struck. That's because
such coatings often fail to bond completely to the metal they
cover. As a result, they seal in moisture or form a barrier that
prevents water from drying out.

PAINTING AN OLDER CAR

When you restore your classic car, you must determine what
kind of paint is on it and how you are going to get it off. To do
a top-rate job, you should remove the old paint completely.

A simple test is to rub lacquer thinner briskly back and forth
on the surface. If your car is coated with lacquer paint, it will
begin to dissolve immediately (*acrylic* lacquer requires a bit of
rubbing before it begins to dissolve). Uncured enamel will begin
to blister or lift but will not dissolve. All other types of paint
(catalyzed enamel — meaning enamel that has had a urethane
catalyst added to it — and base-coat/clear-coat) will remain
unaffected.

You need to know what type of paint is on the car so you
can coat it with a compatible finish. For example, the solvents

in lacquer paint can penetrate uncured enamel and cause lifting; therefore, lacquer should not be applied over enamel.

If the paint is cracked, or if body contours or special lines are difficult to distinguish, the paint will have to come completely off. The same is true for paint that is lifting on corners or sharp curves.

There are three basic ways to remove paint from a car: by disassembling it and having the metal panels submerged in chemicals, by applying chemicals manually and then scraping the paint off, or by sandblasting. Only hand-stripping is practical, unless you are completely rebuilding a car. Then you are going to take it apart and rewire and reupholster it anyway. In that case the submersion treatment would probably be best.

Sandblasting is not practical for most jobs because, on one hand, commercial sandblasters are usually too powerful for use on automotive sheet metal; they quickly overheat and warp the metal. On the other hand, home units are not powerful enough to strip the car efficiently. They require just about as much time as hand-stripping. Furthermore, sandblasting hardens metal, contributing to stress and fatigue. On an older car, this can cause real problems — during later body work, for example.

Before hand-stripping your car, remove all the trim you can (including rubber parts). Then mask the windows (including weatherstripping) and tape over body holes, like those left from the trim, from the inside. Plan on sanding — not stripping — the paint from door and trunk seams, corners and hard-to-reach areas.

Stripping a car by hand is a long and nasty job. Wear goggles, a painter's respirator, heavy rubber gloves and long sleeves. And work in a well-ventilated area — preferably outdoors but not in direct sunlight. Keep a water hose handy.

To strip a full-size car, you will need about 2-4 gallons of stripper. Buy a brand intended for use on automobiles — not furniture or anything else except maybe airplanes. Follow the instructions on the label to the letter, including the guidelines on the number of applications, sitting time, and work temperatures. Don't take shortcuts. And don't be careless. These chemicals are highly toxic.

A sheet of polyethylene laid across the surface after the chemicals are applied will keep them from evaporating too quickly. It will also intensify the stripping process.

As I said before, don't try to strip tight spots; sand them instead. Once you have most of the paint off, wet the car down with water and scrub the remaining patches with steel wool. Then wash the car thoroughly and apply a metal conditioner. Be prepared to prime the car immediately. Bare metal should never be left overnight because it will begin to corrode at once (see Chapter 6 for details on rust inhibitors to protect bare metal).

Submersion. Professional stripping means having the body submerged in rust- and paint-removing chemicals. The companies that do such work, like Redi Strip of Allentown, Pennsylvania, typically dip rusty panels or entire auto bodies in huge vats of alkaline solution. Most also apply a special rinse after dipping. The cost to have an entire auto body dipped and rinsed will generally range from $500 to $2,000, depending on the age of the car and the extent of the rust afflicting it.

General preparation. If the paint on the car is intact, you can probably get by with a thorough sanding before you prime and paint. But give it a top-to-bottom inspection first, to check for rust.

The older the car, the more likely it is to have pockets of body filler hidden beneath the paint. If there are irregular contours or misshapen areas on the body, it's a pretty safe bet they contain filler. If it is plastic filler — not aluminum or metal-based filler — and if it has been there for several years, it has probably begun to shrink and crumble.

There may be other vulnerable areas beneath the paint on an older car, too. For these reasons, it's best to get as much of the old paint off as possible. Then, after the car has been stripped of old paint, grease and rust, all plastic filler must be removed. Use a grinder with a 24-grit open-coat disc for the job.

Believe it or not, there are a few cars on the road that were brush-painted. Such cars require thorough sanding at least, preferably stripping. If you choose to sand the brush marks off, an orbital sander does the best and fastest job.

SPECIAL TREATMENT FOR METALS

When you work on early-model cars, you are likely to encounter, or use, more than one type of metal. For example, aluminum body panels are becoming much more common in body repair. Aluminum is lighter in weight and more resistant to corrosion. In older cars, as I said above, the body panels will often be steel, and the trim and hardware stainless steel, chrome or aluminum. So it is wise to know basic treatments for the most common metals in body repair.

Aluminum. Aluminum is a "softer" metal than steel. When it is exposed to high temperatures, it reaches a threshold rather suddenly and becomes distorted. It reacts similarly under other stresses. Because the metal is so sensitive, you should treat it with special care when you pound, weld or grind its surface.

Follow the guidelines below to prepare aluminum for primer or paint finishes:

1. Clean the metal with solvents. Make sure all dirt and grease are gone.

2. If the surface of the metal is oxidized — that is, if it is spattered with a light coating of corrosion — sand it gently.

3. Buy a metal conditioner made specifically for aluminum, and mix it according to the instructions on the label. Use it to clean the surface with very fine steel wool.

4. Dry the surface with a clean cloth.

Follow the same guidelines for galvanized steel, zinc die casting, zinc alloys, copper alloys and other similar metals. The only difference will be the conditioner itself. Buy one specially formulated for the metal under repair. Treat anodized aluminum parts, such as grilles and bezels, as chrome.

Chrome. Clean chrome parts thoroughly with soap and water. Try to avoid abrasives. However, if you must use one, make sure the metal is completely free of dust and particles, and use a clean cloth to apply it. The chrome on chrome-plated parts is less than 1/1000 inch thick, and perforates easily. Corrosion sets in quickly once the chrome is penetrated. (The only way to repair corroded chrome is to have it replated.)

Stainless steel. Clean stainless steel with an automotive detergent and a soft cloth or sponge. If it is especially dirty, however, you can probably get by with a stiff brush. Stainless steel is relatively sturdy and rust-resistant. Abrasive polishes are OK, but usually unnecessary.

Simulated chrome and other "plastics." Clean simulated chrome with soap and water.

Chapter 6
Chemicals & Coatings

This chapter describes a number of different chemicals and coatings for metal, among them rust removers, converters, and inhibitors; oils; and primers and paints. Most of the products have been borrowed from applications in other fields, such as marine technology and industry. Marine and industrial products for metal protection are some of the most durable in the world. Unfortunately, some may also be lethal for home use (see below).

Although most of the chemicals described in this chapter are easy to find in the Northeastern United States, others may be more common elsewhere in the country. If so, use the information provided in this chapter as a framework for your own research into other products. In other words, ask the questions necessary to get the kind of information included here.

The products in this chapter meet one or more of the following criteria: they are highly effective, available to the public, safe for home use and storage, and/or easily disposable.

For the most part, I have excluded highly toxic products from these pages. A few remain, however, because they are effective and their manufacturers make them readily available to the public. Other products were excluded because their manufacturers failed to provide adequate information on them, despite repeated requests.

Always read the safety information on product labels before using any of these chemicals. Better yet, request a material safety data sheet from the manufacturer — for any chemical or paint you use.

Safety. Manufacturers are required by law to provide the industrial labor force and consumers with material safety data sheets, which list information on the manufacturer, hazardous ingredients and associated health risks, storage, flammability and disposal, among other things.

The primary purpose of an MSD sheet is to provide industrial workers with the information necessary to protect themselves from possible hazards. Therefore, the information on these sheets is usually presented in terms of long-term exposure risks and thresholds. Nevertheless, if you have questions about the safety of any chemical and can't seem to get a straight answer from a salesperson, call the manufacturer and request an MSD sheet.

Below are some abbreviations you will encounter on the sheets and later in this chapter. Some of the governmental bodies, like OSHA, certify protective gear necessary during the application of certain chemicals.

OSHA — the Occupational Safety and Health Administration.

NIOSH — National Institute of Occupational Safety and Health (Department of Health and Human Services).

MSHA — Mine Safety and Health Administration (Department of Labor).

TLV — Threshold Limit Value.

TWA — Time-Weighted Average. For example, the TLV/ TWA for a particular product would be the average concentration to which *nearly* all workers may be repeatedly exposed day after day without adverse effects. Remember, most workers are employed eight hours a day, 40 hours a week. Consequently, they are exposed to much higher concentrations than the average person coating one car at home.

PEL — Permissible Exposure Limit.

ppm — parts per million.

mppcf — million particles per cubic foot.

Flash point — the temperature at which a liquid gives off sufficient vapor to form a flammable or ignitable mixture with air.

When you request an MSD sheet, ask for the most recent one on the product so you'll have the latest information on the health risks of certain chemicals. For example, of several different MSD sheets for the same product, which contains isocyanates, only the latest states that grinding or sanding may be hazardous to health.

In using any chemical, always err on the side of caution. Many MSD sheets and product labels give different recommendations for different work environments. Sometimes the wording is quite vague. For example, a label may say that only a face mask is necessary in areas with "good ventilation," whereas a respirator is required otherwise. Choose the most conservative approach if you have any doubts about the safety of your work area.

And don't be fooled by product literature that features smiling workers spraying paint with no protective gear at all. Read the label. The vast majority of anti-rust chemicals require goggles and gloves as a *minimum* of protection.

Preliminary inspections. Before you choose one of the following chemicals, inspect the surface to be treated. Determine, among other things, where the corrosion originates. Is it inside-out or outside-in corrosion?

If the rust originates on the outer surface of the car, you must choose a chemical that can be topcoated with an automotive finish. If rust has struck both sides of the metal, then the chemical you choose must be applied to both sides.

However, if the metal is so weakened by rust that it crumbles or breaks under pressure, no chemical will save it! Instead, the afflicted metal should be cut out and patched with a sound, healthy replacement. Then less severe corrosion surrounding the patch can be treated with a special chemical.

See Appendix 2 for the names and addresses of product manufacturers listed in this chapter and for additional information on the chemicals themselves.

RUST CONVERTERS

Converters (also called neutralizers) chemically transform existing rust to a stable compound similar to fersoferric oxide, an inert form of iron oxide. Most converters also leave some type of polymer coating to prevent new rust from forming.

One of the main advantages of rust converters is that little surface preparation is required for the treatment of *lightly* rusted surfaces. Converters are recommended for use in hidden areas, like trunks, the engine compartment and the underbody, but they should not be sprayed into enclosed box sections or rocker panels unless the interior surface is only lightly rusted — and it is usually difficult, if not impossible, to determine the extent of corrosion in these sections. Other coatings, such as oil or Oxi-Solv (described later in this chapter) are better for such areas.

Similarly, since they do not smooth moderately rusty surfaces to match the surrounding area, converters may be unsuitable for metal that will be coated with an automotive finish. Moreover, some converters are not sandable wet. However, you may use a converter as a quick *preliminary* touch-up for painted surfaces that are chipped and beginning to rust. A converter will protect the metal until it can be touched-up properly. (As I explained earlier, the metal at the perimeter of a rust spot is especially volatile. Consequently, if rust spots are left untreated, corrosion will creep under the paint coating.)

Rust converters are generally milky whitish-blue in color but turn black after treatment. They are applied directly to lightly rusted surfaces. Two to three coats of a rust converter should generally last at least three years. This section covers five converters: Extend, Rust Reformer, Neutra Rust, Trustan, and Rust Avenger. Other data for these products are given in Appendix 2.

Surface preparation. While most of the manufacturers of converters claim their chemicals are more effective on smoother surfaces, they don't recommend the removal of all rust — only loose or flaking rust. In fact, if a rusty surface is cleaned to completely bare metal prior to application, some companies recommend that it be allowed to re-rust.

To hasten the re-rusting process, use this simple formula. Mix 1 tablespoon of vinegar and 1 tablespoon of bleach in 8 ounces of water. Soak a rag in the solution and lay it against the surface to be rusted. It should re-rust within half an hour. Wash the solution from the metal before applying a converter, however. Also remove any oil or salt and loose paint.

Rust converters should not be applied to heavily rusted metal. For such surfaces, the necessary preparation would be just as involved as preparing for a fish-oil primer: all loose or flaking rust would have to be removed with a wire brush or sandblaster. In fact, one major paint company claims that its fish-oil primer (described below) and a coat of paint will hold up better in salt spray tests than its own rust converter with a coat of paint.

Temperature ranges. The manufacturers' recommended application temperatures generally fall between 50 degrees and

90 degrees Fahrenheit. At colder temperatures, the products react more slowly.

All converters must be protected from freezing until they are applied, although most can survive a few freeze-thaw cycles. However, if a converter appears thick or lumpy instead of watery after a vigorous shaking or thorough mixing, replace it.

The limits on temperature after the converters are dry range from -35 degrees Fahrenheit to 250 degrees for Extend and from -4 to 300 degrees for Neutra Rust (although Neutra Rust will begin to degrade at 250 degrees). These figures are for freshly treated surfaces. I would not be surprised if these ranges narrow with age. (See Appendix 2 for a complete breakdown of temperature ranges.)

It is important to choose a chemical that will withstand the extremes to which your car might be exposed. For example, while I rarely encounter weather below zero at home, I often see temperatures near -20 degrees 200 miles to the north.

Converters are more resistant to weather when they are coated with paint. Some will withstand salt spray seven times longer with a topcoat. Trustan *must* be coated with primer and paint (or another two-part coating system) because it leaves no protective latex coating. When it is topcoated, however, it has "no practical lower limit" and can withstand temperatures up to 500 degrees Fahrenheit. This makes Trustan my choice for metal near exhaust pipes and for cars exposed to extremely cold temperatures. You would not want a coating to crack in the winter, leaving the metal vulnerable to corrosion.

Drying time. Drying time between coats (often called "recoat time") ranges from 30 minutes to more than three hours (see Appendix 2).

Topcoats. If you don't plan to topcoat the converter with paint, make sure you use a converter that leaves a self-protecting film after two to three coats. While most converters can be topcoated without primer, Rust-Oleum recommends that a primer be included for optimal protection.

Health hazards. Most of the converters are labeled "nontoxic," probably under some federal guideline. But all are skin or eye irritants (or both) and contain a small amount of acid (1-2 percent). Extend aerosol is also toxic by ingestion, inhalation, and absorption (through the skin).

Although converters are less toxic than other means of treating rust, you should still wear goggles and rubber gloves when you apply them.

Converters must be properly disposed of after use. They cannot be poured back into the original container or they will contaminate the unused chemicals. 3M's Rust Avenger in the pen-shaped applicator partially eliminates this problem. This design allows for convenient touch-ups of scratches and small chips, but is not suitable for larger areas. However, Rust Avenger is also available in a squeeze bottle (as part of a kit), as is Rust Reformer. Both can be squeezed directly onto rusted areas. But don't drip these products on painted surfaces — and wipe them off immediately if you do.

Most converters are water-based and nonflammable. However, Extend aerosol is extremely flammable. As I have said before, always check the product label for warnings. Check the label even if you have used the product before. A product and its warnings may change, even though its manufacturers may retain the same label design.

Other limitations. Converters are not necessarily the magic potions their advertisers would have you believe. For example, the label of Rust-Oleum's Rust Reformer carries a claim that the chemical works "in one step," without sanding. However, the technical data sheet for the same product says the surface must be scraped or brushed to remove loose rust or scale and any deteriorated coatings.

Similarly, the makers of Neutra Rust claim that the limited surface preparation required for their product makes it superior to other rust-removal methods, which "require expensive labor in intensive preparation of all surfaces." However, Neutra Rust's "limited" preparation requirements include the removal of loose scale, flaky rust, all old paint and dirt.

So read the product labels carefully before you buy a converter. And inspect the rusted metal completely so you will know what type of chemical or special coating will work best.

Rust converters are available at home centers and at marine supply, paint, hardware and auto parts stores.

RUST REMOVERS

Although rust removers may be mistakenly categorized with the converters, they are very different. They *dissolve* rust; converters change rust to an inert form of iron oxide. Cortec 420, 421, 422 and 423 are rust removers. So are Oxi-Solv and Rust-A-Hoy. All are sprayed or brushed on metal and left for different intervals before rinsing. Small rusty metal parts can also be dipped in the chemicals.

What distinguishes these products from one another is their safety — both during use and afterward. Cortec 422/423 and Oxi-Solv stand out because they are exceptionally safe for human use and take a minimal toll on the environment. On the other hand, Rust-A-Hoy and Cortec 420/421 contain phosphoric acid, as do many rust jellies and other products available in hardware stores. Phosphoric acid burns the skin and eyes.

Oxi-Solv. Unlike rust converters, Oxi-Solv can be reused, has an indefinite shelf life, and is biodegradable. And once it is dry, it has no lower temperature limit. It also covers two to 10 times the area that equal portions of rust converters cover. Moreover, Oxi-Solv is nontoxic and nonflammable.

Oxi-Solv leaves a microcrystalline zinc phosphate coating, which makes for excellent paint adhesion (body fillers adhere better, too). The zinc phosphate surface is very smooth, ideal for exterior paint. If it is soaked with Oxi-Solv for several hours, a metal surface can withstand several months of outdoor storage. (The manufacturer recommends a minimum soaking interval of 15 minutes. For heavy rust, two hours is advised. For deeply pitted heavy rust, however, eight hours is optimal.)

In laboratory testing, Oxi-Solv can withstand 500 to 1,000 hours of salt spray when it is topcoated. One hundred hours of salt-spray testing is said to equal one year of normal outdoor exposure, although there is no universal lab test that duplicates real-world exposure for automobiles.

How fast Oxi-Solv works is dependent on the temperature and the extent of corrosion. The manufacturer recommends application temperatures between 60 and 105 degrees Fahrenheit and says 85-100 degrees is optimal. Prior to application, the surface should be free of grease, oil and wax.

During treatment, the metal surface must be soaked, or kept moist with Oxi-Solv. For the interiors of rocker panels, Thick and Heavy Metal Conditioner — a thickened version of Oxi-Solv — is preferable because it adheres well to vertical surfaces.

Any gun capable of spraying rustproofing solutions will work with Oxi-Solv. For exterior surfaces, a Rochester Schutz gun (a trademark of 3M Corp.) is recommended.

Following application, the metal should be rinsed with a mild detergent and water and wiped dry. To dry the interiors of rocker panels, you can use a vacuum cleaner set to blow, aiming it through the stamping holes. Repeat the process above if the metal is still not rust-free.

Any Oxi-Solv that drips out of drain holes (prior to rinsing) can be strained to remove contamination and reused. If Oxi-Solv freezes, simply defrost it and mix it well. Eye protection is recommended during application. Gloves are only necessary for people with sensitive skin or abrasions and cuts.

To treat inaccessible inner body cavities with Oxi-Solv, follow the procedure described in Chapter 3 ("Rustproofing Your Own Car"). After the proper interval, spray a mild detergent and water into the cavities in the same manner. Treat only those sections that have drain holes. After the rust is dissolved, a coating of penetrating oil (see below) is necessary to keep the area free from further corrosion. Park your car on an incline to allow for proper drainage (or jack one end up and then the other, or simply drive through hills for a while). Use a magnet at the stamping holes to check for loose rust, which could clog drainage if it doesn't dissolve completely. During treatment itself, you can plug the drain holes temporarily to allow the Oxi-Solv to soak the most vulnerable parts of rocker panel and doors. If you have silt and sediment built up in the bottom of a door or panel, try to remove it before applying any Oxi-Solv or other chemicals or oil.

Cortec 420 and 421. According to their manufacturer, these chemicals dissolve rust and condition the metal surface for subsequent coatings. They also inhibit rust. The primary difference between the two chemicals is that Cortec 421 is a gel, recommended for vertical and underbody surfaces because of its greater adhesion. Cortec 420 comes in liquid or aerosol form. It can even be used as a dip for small parts.

Both products contain phosphoric acid. And both are formulated for industrial — not automotive — use. So they may have an adverse effect on automotive primers and paints already on the car (as will any acid).

Depending on the seriousness of the corrosion, Cortec 420 and 421 take anywhere from 10-20 minutes to five hours to react. All heavy or loose rust must be brushed off the surface prior to application.

While Cortec 420 can be sprayed into enclosed sections, it is not recommended for such areas unless proper drainage is ensured — and unless the section can be flushed with water afterward.

After any surface is treated with either chemical, it must be rinsed thoroughly with water. Then it can be primed and painted with any *commercial* topcoat — including automotive primers and paints. Without a topcoat, the products will protect bare metal for up to four weeks *indoors*.

Health hazards. Cortec 420 and 421 can cause severe skin and eye irritation. They can also irritate the respiratory and gastrointestinal tracts if they are inhaled or swallowed. Because they are so hazardous, goggles, gloves and a respirator should be worn during use.

Shortcomings. One of the primary drawbacks of Cortec 420 and 421 is their questionable safety for home use. Moreover, although their manufacturer calls them rust removers and *inhibitors*, they still require a primer and topcoat to be very effective. Otherwise, they protect metal surfaces for only four weeks. Moreover, because they contain acid, they will eat away healthy metal if they are left on too long.

Purchasing information. Cortec chemicals are available directly from the manufacturer only, and the minimum order is $150, a fact that complicates individual, home use. However, members of car clubs and other automotive organizations could feasibly buy as a group.

Cortec 422 and 423. These chemicals were only recently put back on the market. According to a Cortec representative, the company has since been swamped with calls from restoration shops and other businesses so the chemicals are available only in limited quantities. Because they were so recently reintroduced, information on the products is still sketchy. However, because they are biodegradable and safe for skin contact, they are probably superior to the other Cortec rust removers.

Like 420, Cortec 422 comes as a liquid or aerosol, and is usually sprayed or brushed on metal. Cortec 423 is a gel. Reaction time, depending on the seriousness of the corrosion to be removed, may be anywhere from 10-15 minutes to five hours. Neither product is recommended for enclosed sections.

The representative says Cortec 422/423 can be coated with most topcoats and primers on the market, including automotive finishes. However, I would still advise you to test for compatibility on another piece of metal first. Follow the directions on the label of the topcoat; they usually call for the removal of all foreign substances from the surface.

Rust-A-Hoy. Like Cortec 420 and 421, Rust-A-Hoy contains phosphoric acid. It can be applied with a spatula or brush, and reacts within 15 minutes to an hour. A second application may be necessary on heavily rusted surfaces. Rust-A-Hoy is not recommended for interior sections.

Because of the acid it contains, Rust-A-Hoy may damage painted surfaces, so apply it carefully. Once it is rinsed, the surface can be coated with any automotive primer and topcoat. Without a topcoat, it will flash-rust.

Rust-A-Hoy is available in marine stores.

Health hazards. Rust-A-Hoy contains phosphoric acid and is a mild skin irritant, and chemical burns are likely if it gets in eyes. According to its manufacturer, it is "slightly toxic" when ingested. Rubber gloves and goggles should be worn during application.

RUST INHIBITORS

Rust inhibitors protect bare metal until it can be processed further. Manufacturers use them to prevent rust during shipping or storage. You may have noticed a film on parts you have purchased for a car — that film was probably an inhibitor.

Some inhibitors are adaptable to automotive applications. For example, an inhibitor might be appropriate for a bare-metal car or part that will be sitting without a topcoat for a number of weeks. The inhibitor will save you the job of removing rust before you paint.

Cortec VCI-319. This product is water-soluble, nontoxic and biodegradable. It is most suitable for cars stripped completely to bare metal — during the restoration process, for example. It leaves a self-healing, protective film. In addition, when it is welded, it emits no noxious fumes. It does wash off easily with water, however, so use it only on garaged cars.

VCI-319 is a synthetic inhibitor formulated as a substitute for oil- or solvent-based compounds, which are extremely flammable. It comes in liquid form and can be sprayed or brushed on metal surfaces. It can also be used as a dip for small metal parts.

One thick coat of VCI-319, which dries in about an hour, should be sufficient to protect most surfaces. Since it is water-based, the product should be topcoated. For that reason, it is not recommended for enclosed sections, such as the interiors of rocker panels.

Surface preparation. Freshly cleaned — preferably sand-blasted — surfaces are optimal (but see the warning on sand-blasting in Chapter 5). Although it can be applied over paint, VCI-319 will react only on chips and scratches — in other words, only surface imperfections will be protected.

Topcoats. VCI-319 does not take the place of a primer, but it does enhance a primer's performance. Its manufacturer claims it is compatible with automotive finishes, but a preliminary test is suggested (see above).

Health hazards. This chemical can irritate skin and eyes; it may also irritate the respiratory tract if it is inhaled. Wear chemical-resistant gloves, goggles, and the proper respirator during application. Since it is biodegradable, VCI-319 can be disposed of easily.

Limitations. VCI-319 can not be applied to aluminum or galvanized steel.

VCI-376. Like Cortec VCI-319, this product is a water-based rust inhibitor that can be sprayed or brushed on metal surfaces. In fact, the two chemicals are similar in most respects: both wash off without a topcoat, both are compatible with most primers and paints, and neither is recommended for enclosed sections. But Cortec VCI-376 differs in the following ways: (1) it can be applied to galvanized steel and aluminum, (2) it is much more durable when it is topcoated, (3) it is toxic, and (4) it can be applied over oily or rusted surfaces.

Health hazards. VCI-376 can cause skin and eye irritation after prolonged exposure. If it is sprayed or atomized, it will create potentially toxic vapors and mists, so a NIOSH-approved respirator is a must during application. Moreover, some of its ingredients can be absorbed through the skin during spraying, so it is important that you dress properly: wear a heavy, long-sleeved shirt, long pants, gloves and protective goggles.

Eastwood Cold Galvanizing Compound is slightly different from the two inhibitors described above. For one thing, it is used primarily to add extra protection under topcoats — not to protect bare metal until it can be topcoated (although it can do that, too). Because it contains zinc, this compound acts on the galvanic principle (hence, its name). That is, in corrosive environments, the zinc coating will give itself up, leaving the steel intact.

Cold Galvanizing Compound comes in both aerosol and liquid forms. Usually, one light coating is sufficient. For abrasion-prone areas, however, thicker coats are recommended (and a topcoat is mandatory). Drying time averages between 20 and 50 minutes for the aerosol, slightly longer for the liquid. The manufacturer also recommends that a sealer be applied before the primer and topcoats.

Like the Cortec inhibitors, this compound is effective only on bare metal, which should be completely rust-free.

Cold Galvanizing Compound works best on new (or exceptionally clean) metal parts and blind areas, such as the interiors of rocker and quarter panels, prior to the application of a topcoat or rustproofing compound. It can also be applied to already-galvanized steel (as a touch-up) and to aluminum panels.

Health hazards. The liquid compound contains what the manufacturer terms "metallic zinc," while the aerosol contains "zinc powder." Consequently, it is best to avoid inhaling either. However, the manufacturer claims on it most recent MSD sheet that Cold Galvanizing Compound requires no more protection than ordinary spray paints: only gloves, long sleeves, and a face mask and respirator in areas of poor ventilation.

Both the liquid and aerosol are flammable. Neither is biodegradable.

OILS

Oil has been used for many years in the Northeast to prevent rust, especially in rural areas. As a result, oil preventives are sometimes called "bumpkin technology." Special oils have also been used in the marine environment to protect ships and nautical equipment. In fact, the U.S. Navy used oils to protect its "moth-balled" ships after World War II. Since a marine environment is much harsher than the conditions a car would encounter, you can expect marine oils to stand up better on land. In Vermont, I recently discovered a sheriff's car with 170,000 miles logged on it and a 20-year-old truck that had been used daily — both with no rust! Both had been treated with ordinary motor oil.

Motor oils. When rust afflicts the underbody of an older car, it usually begins beneath the floors, where pebbles and other debris thrown up by the tires have nicked the car's protective coating. However, wherever there are oil leaks — around the transmission tunnel, for example — the metal is rust-free. That's because the oil keeps water away from the metal.

In most treatments, regular motor oil is sprayed inside rocker panels and doors, under the hood and on the underbody. Holes are drilled for access to rocker panels and box sections. After treatment, these holes are plugged.

The cost for an oil treatment is usually $25-30. Cars are normally treated twice a year — in the fall and spring.

If you want to treat your car yourself, buy a long rustproofing wand and an assortment of spray nozzles (see Appendix 1), and rent an air compressor. Follow the application procedures outlined in Chapter 3 in the section on "Rustproofing Your Own Car." Plugs for any access holes you drill are available in auto parts stores. Avoid rubber parts, like bushings, when you apply any oil.

Please note, however, that motor oil is not advised for interior rocker panels (or the interiors of other enclosed sections) where moisture or dirt, or both, may have accumulated. Although motor oil does form a barrier against moisture, it is quite permeable to oxygen. So if motor oil is sprayed over a surface that is already moist, corrosion will continue to spread as oxygen works its way in.

If you aren't sure about the condition of interior sections, rinse and dry them thoroughly before coating them with motor oil.

Marine oils. Regular motor oil is used in most car treatments. But Consol Clear, a penetrating oil that can be applied to severely rusted steel, does an exceptional job of protecting metal. This oil is normally used in marine applications, and it never dries, preventing rust as it expels trapped air *and* moisture. Consol Clear is available in three grades: light (M1), medium (M2), and heavy (M4). The light oil has a thin enamel consistency. The heavy oil is similar to oil-based house paint. The light oil penetrates the surface quickest and loosens rust, while the heavy oil offers lasting protection (it stays on the surface longer). In fact, the heavy oil will protect metal on the deck of a ship for up to a year. It should work three times longer in rocker panels and doors.

Consol Clear has a number of obvious advantages over regular motor oil. For one thing, motor oil must be reapplied every six months, while Consol Clear M4 lasts up to three years. (You'll know to re-treat the surface when it no longer appears oily.) Second, Consol Clear can displace moisture already present in enclosed sections, while regular motor oil cannot.

Sample application. For lightly rusted rocker panels (where there is no loose or flaking rust), apply Consol Clear's M1 oil, allowing one to two weeks for it to take full effect. Then spray the heavy oil (M4) into the rocker panels.

For heavily rusted rocker panels or enclosed box sections, you would do better to use Oxi-Solv (instead of the M1 oil) because it dissolves rust completely. Otherwise you could end up with loosened rust and scale blocking the drain holes. (If this happens to you, you might be able to remove the rust by using a magnet to pull it through the stamping holes.) If you use Oxi-Solv, follow the recommended soaking times on the label and rinse the treated surfaces well before you apply M4.

Other bonuses. Consol Clear has a very light oily odor. If it drips on painted surfaces, it will not harm them. It can be sprayed on the underside of the floors to protect chipped and rusted areas. According to the manufacturer, it can also be coated with conventional enamel or two-part epoxy paints, making a thorough degreasing unnecessary.

Health hazards. According to its manufacturer Consol Clear is a nonhazardous material under OSHA guidelines. Nevertheless, you should wear goggles and gloves during use and a mask when spraying.

PRIMERS AND PAINTS

Fish-oil primers. Regular automotive primers are sprayed on bare, rust-free metal to protect it from moisture and pollution. They are normally coated with lacquer or enamel paint. Fish-oil primers, on the other hand, are used primarily in areas where complete removal of the rust is difficult, or too costly from a labor standpoint — on wrought iron fences and bridges, for example, and in industry. But they are suitable for a number of automotive applications as well. For one thing, they work on both rust-free and lightly rusted metal by eliminating air and moisture. Their primary drawback is that they cannot be used on exterior surfaces because their solvents will eat through any automotive paints applied on top of them and because the rusty surfaces fish-oil primers are applied to are too rough for glossy finishes. But they are good for floors (inside or out), trunk interiors and other hidden areas. (Fish-oil primers can be covered with compatible topcoats, but such finishes are a far cry from the high gloss or color of automotive paints.)

Rust-Oleum's 7769 Rusty Metal Primer is also fish-oil-based. Available in liquid or aerosol, it is intended for use on sound rusted metal that still appears heavily rusted (reddish brown) or pitted after wire brushing. A rusted metal is sound when it bears no flakes or scale after basic surface preparation. Although Rusty Metal Primer can be topcoated, it isn't compatible with automotive finishes.

Rusty Metal Primer should not be applied to clean metal. Use 7773 Clean Metal Primer instead (also by Rust-Oleum). See Appendix 2 for more information on these products.

The Contact Paint & Chemical Corp. (maker of Consol Clear) manufactures a series of primers and paints for industrial use under the name Lasting Paints Inc. Its Red Penetrating Primer (FSC-1451) is fish-oil based. It is also exceptionally durable and is recommended for use in marine and other harsh environments. However, it contains 7.46 percent zinc chromate and requires an air-line respirator in poorly ventilated areas.

This primer is more effective with a topcoat.

Industrial paints. There are also paints that can be applied over *light* rust. They are usually called "bridge" or "preconstruction" paints. Because they are formulated for industrial use, they can hold up even in severe environments. Unfortunately, they are not suitable for exterior surfaces. Like fish-oil primers, they work best on floors, trunk interiors and other hidden areas. While they can be applied to rusted metal, the surface should be free of all loose scale and flaking rust and paint.

Two-part epoxy paints are one example. They are used primarily for bridges and storage tanks and consumer versions are generally available in large paint stores. However, in my community I have been able to find them only in white and pastel shades. Consol Aluminum Epoxy Mastic 2 (FSC 901 and FSC 902) is a two-part catalyzed system. According to the manufacturer, it is extremely durable and adheres well. It also requires limited surface preparation and resists salt spray. Its main drawback for home use is that it is catalyzed. That means it contains special chemicals to ensure rapid curing without high temperatures. Consequently, the overspray of such paint is extremely carcinogenic; I don't recommend it unless you have a professionally equipped spray booth with excellent ventilation, a lot of spraying experience, and wear a respirator with a separate air supply (read the product label for specifics). Chemical-resistant gloves and goggles are a must. In fact, it is probably best to brush on this paint because its overspray is so toxic.

Someone recently told me about a 20-year-old Volkswagen that had had its floors painted inside and out with industrial paint while it was still new. The car was driven in Michigan, one of the "Salt Belt" states, but never developed floor rot. Of course, today's industrial paints lack lead — a key ingredient in the paints of 20 years ago (and a toxic one, too) — which may render them less effective. But other technological advances (some of which are almost as toxic as lead-based paints) have come along in the meantime.

Other coatings. Take POR-15, for example. It is described by its manufacturer as a "paint-like substance" that can be applied directly to rusted metal. POR-15 is "anhydrous," which means it is formulated without water. Because of that, its manufacturer claims, it is *strengthened* by moisture. Consequently, it dries faster on humid or rainy days and works especially well on metal surfaces with moisture pockets.

POR-15 is available in black (for frames), silver and in a clear primer. Among the types of metal surfaces it will protect are fenders, gas tanks, bumpers, doors, trunks, floorboards, battery boxes, engine compartments, and railings. Moreover, it can be topcoated with regular automotive finishes. It can also be sprayed and sanded. According to the manufacturer, POR-15 works well on surfaces that have been treated with Oxi-Solv. However, it should not be applied over metal that has been treated with a rust converter or metal "prep" or bonding may be hindered. POR-15 should be topcoated as soon as it is no longer tacky, usually within four to six hours.

Once it is dry, POR-15 resists abrasion and expands and contracts with the metal surface. It can be painted directly over rusty metal as long as loose and flaking rust have been removed.

While POR-15 is available to the general public, it contains hazardous ingredients similar to those in some industrial paints — namely, isocyanate prepolymer based on diphenylmethane diisocyanate. A respirator approved by NIOSH/MSHA should be used during application.

POR-15 is a durable coating, but it will break down when exposed to ultraviolet light. Consequently, it must be topcoated.

Health hazards. POR-15 contains isocyanate prepolymer, as mentioned above. So wear gloves, goggles or a face shield, and an air-line respirator when spraying. In areas of poor ventilation, a respirator may be necessary even if the product is being applied by brush.

Once any amount of POR-15 is poured out of the original container, don't return it, even if it remains unused. In addition, be sure to cover the original container with plastic wrap before putting the lid back on, or you may never get it off again.

POR-15 is a product of Stan Coleman Inc.

Other industrial coatings. Other primers and paints are available from Lasting Paints. Some can be applied to rusty metal surfaces that have been wire-brushed and treated with Consol Clear M1 penetrating oil.

Other tips. As a general rule, you should avoid any primer or paint that contains zinc chromate, a suspected carcinogen. Check safety data periodically because paint formulations can change in the course of a few months.

Chapter 7
Choosing a Body Shop

You may have no desire to make your own rust repairs. That's OK. In fact, if you have a real aversion to doing the work yourself, it's best to admit it and find a body shop to do the job. But you should still make a serious effort to learn what the work entails and to understand the steps involved. That way, you'll have better luck choosing a shop to suit your needs. You'll know what questions to ask. More important, you'll understand the answers.

Before you start looking for a qualified body shop, note the following:

- The year, make and model of your car.

- The type of construction it is. Specifically, is it one of the new unibody cars or an older body-on-frame design? (But be aware that some Japanese and European carmakers have used the unibody design since the 1970s.)

- The general extent of the damage that needs repair. For example, are only the rocker panels rusted, or does the corrosion extend into the rear quarter panels, etc.? If the damage is the result of a collision, which part of the car was actually struck by the other vehicle or object? How far did the force travel through the body of the car?

- What your insurance policy will cover.

These are all things you will need to relay to the body shops you "investigate" before you begin posing specific questions.

Don't settle on any shop, even one that is highly recommended to you, without explaining the work you need done. Then ask questions that will help you determine how well that shop will perform the job. And by all means, let the mechanics look over as much of your car as they want. First, however, you should understand a few things about new-car construction and the modern service world.

A few years back, when building a car meant bolting a body onto a frame, cars were much easier to repair. Damaged body panels were pulled and straightened or simply cut out, and replacement parts were welded on. When a car was involved in a serious accident, its frame — the structural foundation — was straightened. Then the replacement panels — mere cosmetic adornments — were attached. Afterward, the front end was realigned. It was that simple.

Nowadays, however, the structural and cosmetic parts of the car are not so clearly defined. Some parts are vital to the structure, some play a purely cosmetic role, and some serve both form and function. If a damaged part is cut out and replaced in the old manner — or even if it is straightened to its original shape — the car's structure may suffer. In fact, such an approach can lead to serious accidents and even death, should another collision occur (see Chapter 11).

The primary concern behind car construction today is safety. Consequently, auto body panels absorb many times the stress they did in the past. The shock of impact gets absorbed by every part of the car except the passenger compartment. As a result, a serious collision can change the size and alignment of body panels drastically.

Once a car has been involved in a serious collision, it is often better to replace the damaged parts because of the incredible stress they have already sustained. And most manufacturers require that any structural panel being replaced be welded at a factory seam. Only minimally damaged areas should be pulled back into their original shape — and then only by properly trained mechanics using the right equipment and factory specifications.

The metal used on today's new cars is much thinner and lighter than the metal used in years gone by. Consequently, conventional gas welding techniques are obsolete in auto body work because they weaken, warp and otherwise distort the metal. Metal inert gas (MIG) welding is preferred.

WHAT MAKES A BODY SHOP GOOD?

One of the best clues to a shop's quality is how modern it is. Has it kept up with technology? Is it prepared to take on unibody cars? Is it well lit? Are at least some of its mechanics trained in the newest straightening and replacement techniques?

Is the shop's frame machine one of the newer types that allows the car to be pulled out to specifications? Does the shop use MIG welding? Does it have the manufacturer's specifications on hand?

These are a few of the more important questions you should ask when you visit any shop, even one that has been recommended to you. And that is how you should begin your search — by getting recommendations from friends or relatives. Even your car dealer can be a good source of referrals.

Just remember that most shops specialize in one area or another. Some prefer collision work; some specialize in foreign cars; some are restoration shops; others are primarily paint shops. Forget the ones that specialize in collision damage if most of the work you need involves restoration, and so on. Narrow your list to three or four shops that claim to have the expertise you need. Then visit their premises. Look around. Talk. Ask questions.

In addition, look for a car club for your model or make of car. You can find listings for car clubs in the classified section of *Road & Track* (probably the most listings) and *Autoweek*, as well as other car magazines. Through these clubs you can find other people who have had good or bad experiences with shops in your area. They may know of a shop that specializes in your model or make. There may also be an "expert" in the club who can identify all the rust-prone areas of your car.

A good body shop should be relatively clean and well organized. It should be prepared for fire emergencies. Extinguishers should be visible and basic safety precautions should be apparent. There should not be open containers of paint or chemicals standing throughout the building (or buildings).

Look for indications that some mechanics are certified by a higher organization or members of professional societies, such as the Society of Collision Repair Specialists and the Better Business Bureau. Also note whether welders are wearing the proper apparel, including face shields, and working in a separate area. Likewise, painters should have their own work space and up-to-date equipment.

Ask what type of welding is being performed and the paint systems the shop prefers. Then look at the equipment on hand. Does it back up the shop's claims? (See the section on welding in Chapter 4.)

Ask what steps the shop follows to paint a car. These steps should include the proper sanding, sealing, masking and priming. If enamel is the preferred paint, ask whether it is catalyzed (to ensure that it cures properly without high temperatures) or whether the shop has access to curing ovens.

Many cars manufactured after 1979 have a base-coat/clearcoat finish. If they require repainting, the job should be performed by skilled technicians using the correct equipment.

Pay attention to any cars being repaired or waiting to be repaired. How badly are they damaged? Also note the general condition of any cars that are waiting to be picked up by the owner.

If your car needs minor — or even major — body work, ask about the shop's repair methods. Ask, for example, how it straightens dents. It should hammer or pull them out or replace the panel completely (if the dent is too big or twisted to be repaired). It should not simply fill the dent with filler. In fact, the shop should not have to apply more than 1/4 inch of filler to any area of the car.

Furthermore, if a panel requires replacement, the shop should be able to justify that replacement. It should also be able to prove that the new panels it installs are of acceptable quality and corrosion resistance. (The use of imitation, or unauthorized, parts may void the manufacturer's warranty; see below.) The new panels should be rustproofed after they are welded in place.

A shop should also be able to guarantee that any rust it encounters is removed completely. Ask how it intends to do this. The corroded metal should be cut out, if it is completely weakened. In the case of surface rust, the metal should be sanded until the rust is eliminated entirely. It should then be treated immediately with metal conditioner.

If a panel is rusting from within, blisters generally appear on the painted surface. It is usually necessary to replace such panels, or other parts, completely. A body shop may be able to patch rusted rocker panels, for example (rather than replace them), but rust may appear a few inches from the repair site in only a few months.

If the metal is generally sound or has only a few small holes, you may want to try one of the chemicals described in Chapter 6 to keep rust from spreading. However, most body shops have probably not heard of these chemicals and will not know how to use them properly; consequently, many will probably be unwilling to guarantee the repair against rust.

Ask which mechanic will be working on your car and what his or her level of training is. If the repairs involve straightening a unibody car and the mechanic has not attended any training sessions in new-car construction, you should ask for somebody else.

If you have read this report thoroughly, you should have a good idea of the questions to ask. Make a list before visiting the shops, if necessary, but make sure you ask the questions.

Request a "damage report" *and* an estimate. They are two different things. A damage report outlines the extent of the damage and plots a strategy of repair. An estimate then assigns each item a cost. An itemized estimate of the labor, parts and materials the repair will require will help you decide between two closely rated shops. But price is not the only consideration. In fact, high quality usually means higher costs. You get what you pay for.

Finally, ask for a written guarantee of the work.

Once you settle on a shop, make a few last-minute checks. For example, contact your local Better Business Bureau and ask if there are any customer complaints on record against the shop you have chosen.

Replacement parts. Many factory warranties explicitly prohibit the use of "unauthorized" replacement parts. At the same time, however, many insurance companies authorize payment only for cheaper replacement parts manufactured by competing concerns. The car owner is caught in the middle of the conflict. Either he pays the difference in the cost of the authorized parts, or he loses warranty protection.

The Certified Automotive Parts Association has arisen out of this conflict. This organization tests replacement parts to ensure that they are strong, corrosion-resistant, and of the proper dimensions. Parts that pass the tests are marked with a yellow sticker bearing the organization's acronym. Nevertheless, before you buy any replacement parts, check with your dealer to make sure your warranty protection is not invalidated.

AFTER THE WORK IS DONE

When you pick up your car, try to inspect it on the premises before driving it home. But make sure you have adequate lighting and that you won't be rushed. Do keep in mind, however, that body shop mechanics cannot wait around all day. They have other work to do. So try to inspect your car thoroughly as quickly as you can.

Here is a list of things to look for:

1. If the car has been painted, in whole or in part, check the trim, windows, weatherstripping, etc., for paint. If you find any, the car was not properly masked.

2. Compare the newly painted areas to the areas still bearing original paint. There should be a close color match. (Keep in mind, however, that older cars may be more difficult to match.) There should not be any dirt particles embedded in the paint.

3. The surface of the repair area should feel smooth. No grinding marks or other nicks or scratches should be visible.

4. The hood, trunk and all doors and windows should open and close easily and in line with body contours.

5. All body lines should be smooth, without protruding moldings or trim or visible high and low spots.

6. The entire automobile should appear to be in the proper alignment.

As you drive your car home, pay close attention to how it "feels" as it moves along the street. Is the drive uneventful and

---

THE ASA

Some of the body shops you visit may be members of the Automotive Service Association, a national trade union of auto service technicians. The group's basic aim is to keep its members informed of the latest procedures so they can continue to turn a profit, says one of the association's Illinois representatives.

While it has investigated customer complaints, the group has no self-disciplinary function. Members can only be expelled if they're convicted of a felony or delinquent in dues payments. The organization does put out a number of publications, however. For more information, write to

Automotive Service Association
1901 Airport Freeway
Bedford, TX 76021

---

smooth? Is the steering straight and steady? Is there any clanking and lurching?

Look at your tires when you pick up the car. They should show even wear. (If they don't, you had a problem prior to the repair.) Check the tires again at 500, 1,000 and 1,500 miles. If an unusual wear pattern develops, your wheels are out of alignment or your unibody was not properly straightened during the repair.

If you encounter any problems once you have paid for the work and brought your car home from the shop, contact the shop immediately. Do not be afraid to complain if you aren't entirely satisfied with the job. The New York State attorney general's office estimates that, of the people who complain about shoddy work to the people or firms responsible, about 55 percent get satisfaction.

If you complain and get no response, contact your local Better Business Bureau, attorney general or consumer affairs office. These agencies have the clout to ensure you a fair deal. If the work is covered by your automobile insurance policy, inform your agent of the problems you encounter.

Chapter 8
Buying a Car

Today's new cars cost an average of $14,000, and used ones average $5,000. That's too much money to hand over to a stranger with no questions asked. But many people continue to walk into dealerships and buy cars with little or no discussion. Later, when a major component breaks down or the underbody begins to rot, these same people have little understanding of their rights as car owners.

In this chapter, I outline some basic procedures that can help you ensure your new car's rust-resistance. Some of the procedures can also help you save money and buy a car that is good all around.

BUYING A NEW CAR

An intelligent shopper is aware of the market. So a man or woman shopping for a new car should know which models are in the highest demand. This will vary from region to region. In the Northeast, for example, many Japanese models are popular. And competition for them is more intense, since import quotas often make it impossible for dealers to keep up with demand.

In such a case, you may have little bargaining power. The dealer can easily sell to the next customer. So take time — before you ever visit a dealer — to decide what kind of car you want and how much you are willing to spend to get it.

One way to make this decision is to do some basic research at home. For example, Consumers Union (see Appendix 1) offers an "Auto Price Service," through which you can order printouts that list the standard and optional equipment for each model on the market. These printouts also list the retail and dealer prices for the equipment. (The dealer price is the amount the dealer pays the manufacturer.) With these printouts, you can determine exactly how much the dealer has paid for the car you want to buy. You can also determine whether you want to buy optional equipment piecemeal or as part of a package (see below).

How new cars are sold. Manufacturers furnish each model line with certain stock components, called "standard equipment." For example, if air conditioning is one of the standard features of the Chevrolet Beretta, every Beretta will come equipped with air conditioning, no matter what. The list of standard equipment varies, however, from one model to another — from the Beretta to the Beretta GT in the Chevrolet family, for example. All standard equipment is included in the basic price of the car.

The manufacturers also make a wide range of other selections available for additional charges. These selections are called "options" and may or may not be included on the cars your dealer sells. Options can be added at any time, however. So if you buy a Beretta without a stereo cassette player, and decide later that you want a stereo cassette player, the dealer or manufacturer will gladly add it on, provided you pay the extra charge.

New-car dealers fill their lots with models they buy each year from the manufacturer. If you buy one of these models from the dealer, you are obligated to pay for any options already added to the car. But don't let a dealer snowball you into paying for an option you don't want unless it was added to the car before you ever looked at it.

Some states prohibit new-car dealers from tying unwanted options to the sale of your new car. In fact, in some states, dealers are required to notify you *before you buy* that you are not required to purchase dealer-added options.

You are always free to order a car from the manufacturer. In such cases, the dealer merely acts as the purchasing agent. When you order a car, you can choose the exact options you want and eliminate the rest.

When the ordered car arrives, inspect it carefully before accepting delivery. Does it have the options you wanted? Does it have options you didn't want? Is the paint coating intact? Is there any damage? If there are any problems, get them corrected immediately, before you pay for the car. The dealer is usually much more responsive at this stage than he will be several months down the road, when he has moved on to other sales.

Know your dealer. Buy a car from a dealer you can trust at the bargaining table, and one who will provide the warrantied service you need *when* you need it. To find a reliable dealer, get recommendations from your friends and colleagues, and check with your local consumer agencies for information on dealers in the area. Sometimes it's better to seek a well-recommended dealer even if you have to pay a little more for the car initially.

Dealer ploys. Unfortunately, some dealers will go to any length to sell you a car at a high price. One ploy that is more and more common today is the grouping of several options in one package. The dealer may try to convince you that it is cheaper to buy four options — like fabric protection, rustproofing, paint sealant, and undercoating — as part of a package than separately. And the dealer may be right about that. But the truth is, you would probably not buy all the options included in the package. You would probably not buy paint sealant and undercoating, for example, if they were not included in a package.

When you look at a car on the lot, pay close attention to the wording used on the window stickers. How are the car's options described on these stickers? Some will undoubtedly be labeled "standard equipment." But are others called "dealer-added" options? If so, you are not obligated to pay for these unless you want them, or unless they cannot be removed from the car. This is where the printouts available from Consumers Union come in handy. They can tell you which options are standard and which are optional. Some dealer-added options include decorative wheel covers, special floor mats, and undercoating.

Another price you may see listed on a sticker is "dealer preparation." Beware. The cost of dealer prep should be included in the basic price, not as an additional charge.

The rustproofing option. Nowadays, most car dealers try to sell rustproofing as part of a package that usually includes undercoating and some kind of protection for the interior of the car. In my opinion, even rustproofing itself is unnecessary, since the manufacturers themselves offer at least six years of protection against rust. In addition, as I explain in detail in Chapter 9, the quality of many rustproofing applications today is questionable. But many new-car dealers will "neglect" to inform you of the factory rust warranty. They would rather sell you several hundred dollars worth of additional rustproofing.

Undercoating is similar. Most cars get undercoated at the factory, as part of the sound-deadening process. The undercoating sold by dealers is usually a heavy spray that tends to create corrosion problems, not prevent them. It also adds needless weight to the car, which reduces fuel-efficiency.

Many people, including dealers, confuse undercoating and sound-deadening. Dealer-applied sound-deadening involves products that are lighter in weight than undercoating and primarily aimed at insulation. However, in my opinion, neither dealer-applied undercoating nor dealer-applied sound-deadening is necessary.

Moreover, the fabric or upholstery protection often included in the rustproofing package is usually high-priced and ineffective.

Recommended options. Some options you might consider are body trim and alloy wheels. Body trim, such as side moldings and bumper guards, can protect your car from dents and dings that can damage the paint coating. But make sure the trim you buy is protective, not nonfunctional. Metal moldings and decorative strips around the windows and doors serve no real purpose. Alloy wheels, on the other hand, resist rust and are lightweight.

New-car damage. At some new-car dealerships, as many as 60 percent of the cars that are sold have had panels repainted. And these are new cars! The fact is, even new cars get damaged, usually during transport to the dealership. But how do you tell which cars have been damaged and repaired, and which are in good condition?

One way is to examine the car's surface closely. Choose a clear, sunny day to conduct your inspection. Don't do it at night or when the sky is cloudy or foggy. Look for variations in the paint tone or sheen. A device called Spot Rot™ can help. It pinpoints areas that have been damaged and repaired, even areas afflicted by rust. This product is described in more detail later in this chapter.

The warranty. The last few years have seen the rise of the "warranty wars" between the major American automakers. Even some foreign automakers — Honda, for example — are entering the foray. At present, Chrysler leads the American pack, with a seven-year warranty on its 1989 models.

Warranties that appear to be generous sell cars. But before you buy any new car, ask to see a copy of the warranty. And read it! (For more information on new-car warranties, see Chapter 10.)

BUYING A USED CAR

With the high price of new cars today — averaging $14,000 — and the added expense of repairing them, it isn't surprising that more and more people are looking to buy used cars or, as many dealers like to call them, "previously owned" cars. In fact, one expert (John Pfanstiehl of Pro Motorcar Products Inc.; see Appendix 1) claims that two out of every three cars sold in the U.S. are used cars, and that most U.S. households have at least one used car.

Chances are, sooner or later, you will buy a used car. The steps I outline below can help you choose a healthy one and avoid costly repairs.

A few basic facts. Did you know that about 20 million cars annually are involved in collisions? That most of these damaged cars are repaired and resold? That an increasing number of cars are "clipped" each year? (This means that a damaged front or rear end is *clipped* from the car; then the good half is welded to the good half of another car. Clipping is illegal in Europe. In the U.S., however, it is preferred by more and more body shops and insurance companies.)

These facts, courtesy Pro Motorcar Products of Clearwater, Florida, highlight many of the risks involved in buying a used car. But these risks can be overcome if you know what to look for when you make your purchase.

Who do you buy from? You have several choices. You can buy from a private party by answering an advertisement in your local newspaper. You can buy from a relative, friend or other acquaintance. You can buy from a used-car dealer. Or you can buy from a new-car dealer. All have advantages and disadvantages.

When you buy from a private party — a stranger, in other words — you are generally able to spend more time going over the car without feeling pressured. On the other hand, a stranger — because he or she is just that, a stranger — may feel no obligation to point out the car's shortcomings or other drawbacks to the purchase (as a friend or relative probably would). Nor will the stranger be as flexible in price.

A used-car dealer may offer a limited warranty and have a wider selection for you to choose from. But many are in business only temporarily. Others knowingly hide serious problems.

And, while you can generally count on a new-car dealer to sell cars that are in good condition (and often to provide a warranty as well), you can also expect to pay top dollar in the process.

It doesn't matter so much who you buy from as what you know about the car. Before you hand over any money, you should have a good idea of the car's history. So decide what you want before you start looking around. Check several different sources. And ask questions. Also inspect any prospective buy thoroughly.

What do you look for? When you examine a car for sale, what do you look for? Well, there are different guidelines for different cars, as well as a list of standard points to check. First, go over the car from top to bottom (and the interior too), looking for general flaws, such as dents and dings, chipped paint, torn upholstery, missing trim, etc. Then go over it again, only this time more closely. Run your hand along the surface, if necessary.

Do you see any places where the car may have been repaired? If so, there may be pockets of body filler. Is the paint even in tone? Or does it vary in shade from one area to another? If it does, the car may have been in a collision. It should be checked for proper alignment and repair.

Is the paint bubbled in spots? Rust may have already begun to spread beneath the surface. Are all the body panels steel? Or are some fiberglass or aluminum? If they are fiberglass, they could camouflage a rusty frame ....

You get my drift. If you know how to read the body of a car, it can tell you much about its life.

Check the underbody, too — or should I say *especially*. Look for corrosion in the wheel wells, around the rocker panels, door edges, box sections, exhaust system, brake lines, etc. Use a flashlight. If the car appears to have been treated with under-coating *recently*, there is a good chance it conceals serious problems.

In the interior, look for signs of water leaks. These include damp floor mats and a musty odor. Lift the floor rugs and check the metal beneath them. If you can't lift the rugs, tap different sections of the floor loudly. Any signs of holes or metal replacement?

Try opening and closing the doors from the inside, and rolling the windows up and down. Try out the radio and the heating/cooling system. Adjust the seats. Honk the horn. Does everything work? Also check the interior of the trunk; look under the mats and at the seams on the lid.

Check for consistency. For example, if the car is fairly new but has badly worn parts, like deteriorating rubber on the pedals, you should ask about the discrepancy. You should do the same if the car is old and has many new parts. See pages 141 and 143 for a complete list of areas to inspect.

Finally, inspect the car on a clear day — never at night or in the rain or fog. Dim light or even the bright lights of a car lot at night can camouflage scratches, chips and other surface imper-fections.

INSPECTING USED CARS

- Check the paint surface for flaws. Rust may develop where there are scratches or chipped paint. Bubbled paint indicates rust that has already begun to spread beneath the surface. And an uneven paint tone may indicate collision damage.

- Inspect areas that get exposed to water, mud and slush. These include the bumpers, wheel wells, front fenders (top and bottom), underbody, door bottoms and rocker panels.

- Examine the underbody thoroughly, including box sections, brake lines, the exhaust system, etc. Is the undercoating intact? Was it applied recently? If so, it may conceal big problems. If it's old and beginning to flake, serious trouble could lie ahead.

- Look at the body panels. Are there any wavy contours or high and low spots? If necessary, run your hand along the panels to feel for hidden repairs.

- Check the engine compartment, including the battery tray, hood lid, and inner wheel wells, for rust.

One device on the market can help with your inspection. It's called Spot Rot™. If you hold it against the surface of a car, it will give a numerical reading that indicates how sound the steel is underneath. If you check different areas of the car in this manner, significant variations in readings should alert you to rust, body filler, nonmetal parts, or other defects that you should ask specific questions about. See Appendix 1 for more details on this product.

What questions do you ask? Before you buy a car, you want to know as much as possible about its history and the reasons it is being sold. Tailor your questions accordingly. Don't hesitate to ask direct questions, either. Why is the car being sold? What kind of treatment has it received? Here is a list of other things you will want to know.

1. Is the factory warranty still in effect? If so, is it transferable? What does the transfer process entail? Is there a fee involved? What is excluded from coverage? (For accurate information on the transferability of a particular model's warranty, you can always contact the manufacturer's zone or district office.)

2. Is there an after-market rustproofing warranty still in effect? If so, what is its duration? Is it transferable? Is there a fee? Does it require regular inspections? What is excluded from coverage? (See Chapter 9 for more information on after-market rustproofing.)

3. Has the car ever been in an accident? If so, what was damaged and when? What did the repair involve?

4. Has the car been driven regularly or has it been sitting? If it hasn't been driven much, why not? If it has been sitting, how long?

5. What was the primary use of the car? Was it for city or long-distance driving?

6. Does the car have any idiosyncracies?

7. Is the seller the original owner?

8. Is the seller a dealer? If so, and you are answering an ad that indicates otherwise, buy elsewhere.

9. Does the car have any mechanical problems?

To be safe, you should ask to see any warranties that are still in effect, and you should make sure the seller is going to fulfill his transfer obligations. Also check the expiration dates of the inspection sticker and the registration. If either is due to lapse in

---

MORE INSPECTION TIPS

- Does the car have any fiberglass panels? Are they in good condition? If they are full of hairline cracks, they could be impossible to repair. Moreover, if the rocker or quarter panels are fiberglass, the car could be dangerous because these are key structural areas on a unibody car. Inspect the metal frame beneath the fiberglass panels. Is it rusted?

- Examine the interior of the car. Inspect the floors for rust and the upholstery for stains or rips. Does the interior hardware (like window handles) function properly? Check the pedals for wear.

- Make sure all body trim is intact. If the trim is metal, check the area around it for corrosion.

- Check the drainage holes. Are they blocked? Is the metal around them corroding?

- Are the headlight and taillight assemblies damp or corroded inside?

- Is the car's inspection sticker due to expire soon? If so, ask about it.

---

the next few days or weeks, the owner could be trying to sell the car because he or she knows it will fail inspection. The owner should be willing to lower the price of such a car.

Fiberglass bodies. If you are thinking of buying a car with a fiberglass body, be sure to check the frame for corrosion. If you have to inspect the car at a lot or in a stranger's driveway, this may seem next to impossible, but it isn't. Simply kneel or lie down and look at the frame side rail behind the front door. Is the steel smooth? Or is it dimpled or pock-marked? It should be smooth.

Find one of the frame's drain holes. Are the edges smooth? They should be. Can you see the full thickness of the metal? You should be able to. If the hole is large enough, insert a finger and feel around. Is there dirt or flaky metal inside? You are bound for trouble if there is.

If the frame has been coated with paint or another substance, rust may have already struck. That's because such coatings often fail to bond completely to the metal they cover. As a result, they seal in moisture or form a barrier that prevents water from drying out.

If the frame is covered by thick undercoating, use a screwdriver to scrape some of it away and poke at the metal underneath. (If the owner won't let you, the undercoating is probably hiding serious defects.) No matter how strong you are, you won't be able to force the screwdriver through good, sound metal. Only rust-weakened metal will cave in.

Also look for patches where metal has been welded onto the frame to cover holes or corrosion.

Checking the mechanical components. If the car passes your inspection, ask the owner if you can take it to a mechanic, or if you can bring a mechanic to the car. (If the owner balks, forget the car.) Tell the mechanic about any questionable areas you noticed, and have him (or her) test-drive the car. If the mechanic finds any items that need repair, get a written estimate of the cost of parts and labor. You can use it to negotiate with the seller — if you are still interested in the car.

Many mechanics will examine cars for a set fee, usually about $25-50. There are even special services set up for this purpose, which usually charge a flat rate per car. But beware! Since these services charge a flat rate for each car they examine, they stand to benefit from prolonging your search.

THE LEMON LAWS

Many states have enacted "lemon laws" to protect consumers from defective automobiles. These laws provide a legal remedy for consumers who buy cars that are "lemons." Where lemon laws exist, there are usually separate versions for new and used cars.

In New York State, for example, a person can choose either a full refund or a comparable replacement vehicle if his or her new car fails to meet the terms of the written warranty and cannot be fixed after a reasonable number of attempts. The law for used cars requires the dealer to provide the buyer with a written warranty. The dealer must also repair — free of charge — or replace used cars that are lemons.

Of course, these laws were written to protect consumers from mechanical problems more than anything else. In addition, the laws are more complex than they appear on the surface. To find out more about your state's lemon laws, contact your attorney general.

Chapter 9
After-Market Rustproofers

Although the automakers have stepped up their anti-rust technology, most new-car dealers continue to push additional rustproofing on their customers. For a fee that ranges from $150 to $550, new-car buyers can have their cars fogged or coated with chemicals that are supposed to retard corrosion. This job is usually done by an after-market rustproofer — a company (apart from the automobile manufacturer itself) that specializes in such procedures.

Ziebart, one of the oldest and most financially successful rustproofers in the U.S., estimated in 1979 that 3,600,000 vehicles received "after-manufacture" rustproofing annually. So you can see that the business of rustproofing was lucrative even 10 years ago.

Under ideal circumstances, the after-market rustproofing process works in the following way. The customer buys a new car — including the rustproofing option. The dealership then has its own personnel treat the car, or farms the job out to an independent company like Ziebart. The person applying the chemicals follows a set of written guidelines and instructions as he or she coats the car. Once the car is treated, it resists corrosion, even in the worst of environments. A warranty is included to protect the customer — just in case .... Unfortunately, the application of after-market rustproofing is rarely this simple.

New-car dealers and rustproofers will tell you — and in some cases, perhaps, even rightly so — that after-market rustproofing can add extra protection that is sorely needed in extremely corrosive environments. Take the Great Lakes region, for example, where acid rain and road salts form a double threat. They also argue that, despite the automakers' recent advances in anti-rust technology, cars remain vulnerable to corrosion.

But in many cases, after-market rustproofing may cause rust, not eliminate it. In fact, some manufacturers advise against the procedure. In a 1988 GM owner's manual, for example, the following warning appears:

> the application of after-manufacture rustproofing is not necessary or required .... In fact, some after-manufacture rustproofing may create a potential environment which reduces the corrosion resistance designed and built in your vehicle .... some after-manufacture rustproofing could result in damage or failure of some electrical or mechanical systems. Accordingly, repairs to correct damage or malfunctions caused by after-manufacture rustproofing are not covered under any of your GM New Vehicle Warranties.

WHEN RUSTPROOFING IS A WASTE

WARRANTY: As long as it has been applied correctly, we stand behind our products 100%, so we encourage you to offer a limited warranty to your customers. However, as most buyers will either trade or sell their vehicles within three years, there is no reason for you to be overly concerned. Even though we firmly believe that we have the best rustproofing coating in the world, it is only as good as the person who applies it. Even though we supply you with complete instructions in this folder, one of your personnel may occasionally miss a fender or panel. Unfortunately, these things do happen, that is why we have carefully inserted a line in our warranty that states, "...this limited warranty covers all areas of passenger vehicles that show evidence of having been coated with DEP RUST-GARD rustproofing material..." Thus, the few customers that might return with rust problems will probably have them in areas that we are not responsible for.

In lots of ten, your price for DEP RUST-GARD kits is only $19.95 each. One kit is more than enough to process a new vehicle. The coating is also available in five-gallon pails and fifteen-gallon drums for volume users.

RETAIL PRICES: There are no set retail prices for our rustproofing. However, most of our dealers charge between $150 and $200. Remember, you can charge as much as you want because you have the vehicle on the premises. Play up the convenience of your customers having the work done before they pick up their new cars. Also remind them that they can simply add the price of the rustproofing to their payments. They will never miss the extra dollars a month, and in many states the law allows you to charge the maximum interest rate for new-car add-ons, thus increasing your profits even more.

These statements were excerpted from the *Application Instructions Booklet* for DEP RUST-GARD (reprinted in *A Tarnished Option — The New York Attorney General's Report on the Rustproofing Option Sold by New-Car Dealers*, NYS Attorney General's Office, June 1981).

There are always people searching for new ways to make easy money and, unfortunately, selling after-market rustproofing at inflated prices is one quick route to revenues. This is not to say that everyone who deals in after-market rustproofing is deceitful. But, in large part, dealers, rustproofers and chemical companies all expect new-car buyers to continue to pay for services without asking too many questions. They also expect most new-car buyers to sell their automobiles before any rust problems have a chance to develop.

In an investigation centering on the rustproofing options sold by new-car dealers, the New York State attorney general found that the average retail price charged for rustproofing was triple the dealer's actual cost. (The findings of this investigation are outlined in *A Tarnished Option*, cited above.) In fact, in 1980 alone, consumers paid more than $11 million for rustproofing that was not rendered or that was rendered inadequately! The attorney general also found that, in many cases, basic car surfaces were not getting treated at all. Moreover, this poor-quality rustproofing was not confined to any particular brand of chemical or to any particular price level.

WHAT ADEQUATE RUSTPROOFING INVOLVES

When any rustproofing concern treats your car, it should have — as part of its standard equipment — detailed guidelines on the make and model. These should be printed and kept as part of a manual for easy access, or diagrammed on microfilm or microfiche. The guidelines should indicate each area to be treated. They should also describe how access to that area is to be gained, and specify the tool to be used. Moreover, a qualified rustproofer should also have the tools necessary to treat all the different types of surfaces, and the person applying the chemicals should have undergone special training.

Every car has basic critical surfaces that must be coated. These include taillight sections, interior trunk panels, quarter panels, pillar posts, the lower firewall, fender panels, the hood, headlights, fender eyebrows, gravel shields, structural frame parts, the entire underbody, rocker panels, wheel wells, the gas tank area, and the interiors of the front and rear bumpers, body panels and doors.

In addition, the rustproofing chemicals themselves should seal out moisture and flow into even the tightest crevices. The sealants must be "self-healing" so scratches and other minor forms of damage that occur later will get automatic coverage. The sealants should also remain flexible and resist abrasion. Finally, the rustproofing chemicals should penetrate seams, welds and cracks.

HOW TO NEGOTIATE A PACKAGE

Most people who buy after-market rustproofing do so when they purchase a new car. And new-car dealers use several methods to add rustproofing to the purchase price. If they can add it quietly, many will choose to do so. In recent years, for example, new-car salesmen have often *not* explained to their customers the extent of the manufacturers' anti-rust measures. Nor have they pointed to the manufacturers' anti-perforation warranties, which are, for the most part, more extensive than any warranty given by independent rustproofers.

The New York State attorney general found that, "When selling rustproofing as a dealer-applied option, new-car salesmen generally do not disclose the existence and nature of the car manufacturer's rust protection or the manufacturer's anti-corrosion warranty .... New car buyers are also not advised that, during the term of the car manufacturer's rust protection warranty, they receive no additional warranty protection from the after-market rustproofing warranty furnished by the dealer."

In addition, many dealers have tied the after-market rustproofing option to other, more practical options on new cars, compelling the buyer to pay for the rustproofing to get them.

When I called a number of dealerships in my area to inquire about the *factory* warranty, I was given distorted information and pressured to buy additional protection. For example, one dealer told me the manufacturer only covered outside-in perforation (the opposite of the truth). Another told me that the factory warranty for rust perforation was only three years (it was actually five). While they were maligning the factory warranty, these dealers went out of their way to extol the virtues of the after-market warranty. And this was only over the phone!

So the first thing you should know, if you are considering after-market rustproofing, is that the federal Magnuson-Moss Act requires new-car dealers to give their customers a copy of the warranty — for any after-market rust treatment being considered — before selling such treatments. (After-market *rustproofing* warranties will be covered in more detail later in this chapter.)

The second thing you should know is that some carmakers (General Motors and Mercedes-Benz, for instance) discourage after-market rustproofing on some models. They argue that the anti-rust technology applied in the factory is incompatible with the chemicals used by after-market rustproofers. In some cases, they also argue, factory technology may be incompatible with the techniques used by the rustproofers to apply those chemicals. Moreover, they say, after-market rustproofing may actually accelerate the corrosion process.

Ask questions. If you decide to buy after-market rustproofing for your new car, ask some basic questions. The most important questions should center on the experience and training of the person who will actually be treating the car. Has he or she done this sort of thing before? (Let's hope so.) Has he or she completed any special training courses? Is there a list of specifications for your particular make and model? What tools will be used?

Many rustproofers drill holes in the automobile to gain access to out-of-the-way areas. Find out whether this will be done with your car. Then ask how the holes will be treated and sealed to protect them from corrosion.

Some experts discourage the drilling of holes. They say that each hole creates a new entry point for rust and can hinder drainage in your car. But others argue that some areas are impossible to reach unless holes are drilled. They say that, as long as the holes are coated with chemicals and sealed properly, there should be no problem. The federal government, by the way, supports the drilling of holes in its list of standards for the rustproofing of federal vehicles (see the box on p. 154).

In addition, ask specifically about the underbody of the car. Will it be treated? How will it be treated? If the salesman mentions "undercoating," make sure he isn't talking about sound-deadening or similar materials. They can actually seal in dirt and moisture and cause rust.

Ask what brand of chemicals will be used, and whether the automobile will be "fogged" or "sprayed." Most experts agree that the spraying method is more durable. Fogging sometimes results in uneven or inadequate coatings.

Finally, call your Better Business Bureau or state attorney general and check on the chemical company (the manufacturer of the rustproofing chemicals). While you are at it, check on the new-car dealer and the rustproofer, too. However, if the business you are asking about is fairly new, it may not have been around long enough to generate any complaints.

When to rustproof. Is it better to rustproof a car when it is brand new, before it has even left the lot? Most experts seem to think so. But a few claim that rustproofing will be more effective

FEDERAL SPECIFICATIONS

You might be interested to know that the federal government, as part of its routine maintenance program for federal vehicles, makes rustproofing a priority. It has adopted a list of standards for the process, which are included, in part, below:

- **Limitations.** Rustproofing will generally be limited to vehicles operated in coastal areas, the snow belt, the Great Lakes region, and areas of high atmospheric pollution.

- **Spray tools.** Spray tools shall be capable of coating a full 360 degrees into all critical areas under high-pressure heads, and be capable of operating through 1/2-inch access holes to penetrate all critical interior areas.

- **Technicians.** Rustproofing technicians must be trained in correct coating techniques by a responsible rustproofing company or by the manufacturer of the equipment used during application.

- **Engineered instruction.** Component illustrations shall be furnished by the vehicle manufacturer and drill locations established by trained technicians.

- **Detail requirements.** The areas listed below shall be properly coated: front, front and rear fender, engine compartment, hood and deck lid, cowl panel, doors, pillars, dog legs, quarter panels, rear lights, trunk compartment and trunk floor extensions, seams and moldings, rocker panels, body floor supports, frames, underbodies, brake lines, rear lift doors, panel and pickup truck rear panels, truck doors, truck bodies and cabs, and truck floors.

if the car has been driven a couple of hundred miles first. They say this allows the body to bend and flex, creating a network of tiny cracks in the paint and primer, primarily along seams and welds. Once this network of cracks has been created, anti-rust chemicals will flow in and protect it. If the chemicals are applied before the car has been driven, these experts claim, many of the cracks will lack protection when they do develop.

Whether or not the car is driven before rustproofing, it should be thoroughly cleaned and dried before any chemicals are applied. Cars get dirty, even during transport to the dealer. If dirt and moisture are trapped between the rustproofing chemicals and the car, the chemicals will not adhere properly and the site may become a corrosion pocket.

There are advantages to having the dealer arrange the rustproofing application (whether the job is performed at the dealership or at another company, like Ziebart). Convenience is one. But you may get better results if you contract a rustproofer yourself, without the dealer's input. For one thing, a rustproofing specialist's whole business centers on rustproofing. New-car dealers have bigger fish to fry. Furthermore, at the plant of a rustproofing specialist, you may have greater access to the technicians themselves, and more leeway in negotiating the terms of a warranty.

What about used cars? A few rustproofing specialists offer plans for used cars. Ziebart comes first to mind. These plans may even include warranties for two- to five-year-old cars that are rust-free at the time of application. But most after-market rustproofing involves new cars — cars three months old or younger.

Negotiation checklist. Before you agree to pay for rustproofing, get the following information:

1. Has the technician been specially trained by the chemical or equipment manufacturer?

2. Do the chemicals to be used include a seam penetrant? (They should.)

3. Will the sealant remain flexible and resist abrasion? Is it self-healing?

4. What tools will be used?

5. Are there updated specifications on hand for each make and model of car treated?

6. Will the car be washed and dried before it is treated?

7. Where will holes be drilled, and how will they be treated and sealed?

Questions concerning the after-market rustproofing *warranty* should also be posed. These will be covered later in this chapter.

After your car is rustproofed. Before you drive away from the dealership or rustproofing specialist, conduct an inspection of your own to ensure that your car was treated thoroughly. Although you may not be an expert on chemical treatments, there are basic checks you can make.

First, the rustproofing chemicals are usually darkly colored and easily visible. If you aren't sure, ask the rustproofer or dealer what the chemicals look like — whether they are transparent or tinted (some may show up only under ultraviolet light). Then go over your vehicle and check basic surfaces for coverage. When you come to holes that were drilled, unseal them (they are usually filled with a plug or cork of some sort) and poke a stick into the enclosed section. If the stick comes out dry, the section was improperly coated — or not coated at all. If there are no holes, and you were under the impression holes would be drilled, then something is seriously amiss.

If you find areas that were coated poorly or neglected completely, you should ask the rustproofer to treat them again.

If the rustproofer balks, contact your local Better Business Bureau or state attorney general.

THE RUSTPROOFING WARRANTY

In 1980 and 1981, the New York State attorney general's office focused on the after-market rustproofing protection sold by new-car dealers in the state. (The attorney general's findings are outlined in *A Tarnished Option,* cited earlier.) The office had been getting more and more complaints from consumers whose cars had begun to develop rust. These consumers were angry because they were finding it difficult or impossible to get warrantied work from the new-car dealers or rustproofers who had sold them the protection. (Other states' offices have conducted similar investigations. Check with the attorney general in your state for further information. Also be aware that, in the years since the investigation described here, many new-car dealers and rustproofers have improved the quality of their services; however, during some basic investigative work in my region, I encountered many of the same problems described by the attorney general.)

The rest of this chapter outlines many of the problems the consumers encountered when they tried to get warrantied service from after-market rustproofers. When I mention *warranty* in this chapter, I mean the rustproofer's *after-market* warranty — not the manufacturer's new-car warranty, unless I state otherwise.

Among the New York State attorney general's findings were the following:

1. The warranties offered by different after-market rustproofing concerns vary significantly. But they all include "limitations or conditions that render them essentially worthless."

2. While the new-car dealers themselves usually provide their customers with the warranties, "they generally do not assume legal liability. Rather, liability is assumed by the rustproofing company chosen by the car dealer .... If the rustproofing company becomes insolvent, the consumer is left without recourse."

3. "[A]ll [after-market rustproofing] warranties limit protection to instances of perforation from inside out and do not protect against rust caused by stone chipping, accidents, damage or paint deterioration."

4. In duration, registration requirements, transferability, inspection requirements, exclusions, filing procedures and remedies, the warranties "differ significantly."

5. "[V]irtually all the warranties exclude implied warranties or limit their duration. None allow recovery for consequential or incidental damages."

The independent rustproofers can afford to offer "lifetime" warranties because so many people will sell their cars and fail to transfer coverage. Many others will be involved in collisions, miss inspection deadlines, and so on. As a result, only a small percentage of the consumers who might actually qualify for warrantied service will ever come forward to seek such service.

Even so, warrantors regularly include several loopholes in their contracts. In its report, the New York State attorney general's office pointed out the 10 most common loopholes. These loopholes are described below. If you are considering after-market rustproofing protection, and the written warranty includes one of these loopholes, ask to have it deleted before you sign the purchase agreement. If the warrantor refuses, buy your protection elsewhere.

WARRANTY LOOPHOLES

One remedy. Many rustproofing warranties across the country protect the warrantor by limiting responsibility to one rust repair — no matter how small — or a refund of the cost of the rustproofing. In most of these cases, it is the warrantor who gets to decide between repair and refund, not the consumer.

As I mentioned earlier, rustproofing concerns bank on the fact that most people sell their cars within three to five years. This is usually too short a period for significant rust to develop — or at least to be noticed. The one-remedy loophole protects the rustproofer in the few instances where a significant rust problem does develop. Significant rust is rust that would cost more than the price of the rustproofing to repair. With such a loophole in the warranty, the rustproofer can opt to refund the initial cost of the job. Meanwhile, the consumer still has a serious rust problem to contend with.

Proper application. Some warranties contain a clause that excludes areas of the car that were not "properly coated," or one that limits coverage to "treated areas." With such a loophole, the warrantor can claim — when rust develops — that damaged surfaces were overlooked or inadequately coated during the rustproofing process. In effect, the warrantor gets off the hook by hiding behind his own employees' negligence.

Registration. Some warrantors require car owners to register their vehicles within a certain period. This means that the consumer must register even after he or she signs a contract with the new-car dealer and pays for the rustproofing. Even after the car is coated, the warranty can be invalidated if the consumer has not filled out the proper registration documents.

Transferability. Many new-car dealers will urge you to buy after-market rustproofing because, they say, it increases your car's resale value. Some warranties allow car owners to transfer coverage when they sell their car, but many do not. Furthermore, many warranties that allow coverage to be transferred do so only within a limited amount of time. For example, Warrantor "A" might allow Bob to transfer coverage to Bill, as long as the proper application is filed within, say, 30 days after Bill buys the car. In many cases, a transfer "fee" is involved as well.

Inspection requirements. Many companies require car owners to return for periodic inspections. Car owners who fail to do so may have their warranties invalidated. And often the inspection requirements are rather rigid or arbitrary. For example, one company (see the chart in Appendix 3) requires that inspections take place within 30 days of the third, sixth and ninth anniversaries of the original rustproofing job. Another requires annual inspections that must take place within 30 days of the original job. Another requires inspections every 18 months.

In addition, upon each inspection, the car owner may be charged a "reasonable" fee for the "removal of dirt" to facilitate inspection.

Accidents. One common loophole protects the warrantor in case the car is involved in a collision. In many cases, if the car owner fails to notify the warrantor immediately and to bring the car in for inspection and re-treatment, coverage may be canceled. It may be canceled even if rust develops in an area unaffected by the collision.

Manufacturing defects. Another way a warrantor may get off the hook is by claiming that the rust was caused by a "manufacturing" or "design" defect. The wording in such loopholes is usually vague — devoid of any coherent criteria for making such decisions.

Pre-existing warranties. Lots of car buyers miss this one. This loophole means that as long as the manufacturer provides anti-rust coverage for the vehicle, the rustproofing specialist is not liable — only the manufacturer. Consequently, if Chrysler's anti-rust warranty is seven years, and the after-market rustproofing warranty is seven years, the rustproofer is not liable at all.

In addition, as I stated earlier, some manufacturers advise against after-market rustproofing and may not cover rust that breaks out on a rustproofed car.

Limited remedies. This loophole is similar to the first one mentioned. It means that, if significant rust develops, the warrantor is only liable for repairs that cost the same as the initial rustproofing — or less. Other variations of this loophole include a ceiling on recovery, or pro-rated refunds of the initial cost of the rustproofing.

Notification. Some warranties require the car owner to notify the warrantor "immediately," "promptly," or "within a reasonable amount of time" when rust is noticed. This sounds fair enough, doesn't it? In reality, however, the vagueness of the wording protects the warrantor, not the car owner. That's because the warrantor gets to decide whether or not notification meets these criteria.

SHOPPING FOR AFTER-MARKET WARRANTIES

Federal law requires new-car dealers to provide customers with a copy of the after-market rustproofing warranty before the customers buy rustproofing protection. Ask for your copy. If you don't get it, call your Better Business Bureau or state attorney general.

Some basic questions. As I said before, the most important variable in after-market rustproofing protection is the technician who applies the chemicals. But even with the best of technicians, you may later encounter rust that needs repair and should be covered under your warranty. When you first consider after-market rustproofing, ask a lot of questions about the warranty itself. Use the samples below as your framework.

1. Does the warranty cover the entire car, from top to bottom? And does it cover both materials and workmanship?

2. Who is the actual warrantor? Is it the dealer, the product manufacturer, or the rustproofing specialist? If one of these should become insolvent, will you still be covered?

3. Does the warranty cover rust damage, or does it just provide for a refund?

4. Does it exclude certain areas of the car from coverage? If so, which ones? Why?

5. Does it exclude areas of the car that were poorly coated or forgotten by the technician? (It shouldn't, of course.)

6. What is the life of the warranty? Does it last as long as you own the car or only a certain number of years or miles? If it's a "lifetime" warranty, does coverage end when you sell the car to someone else?

7. Is the warranty transferable? Under what conditions? Is there a transfer fee?

8. Do the dealer and the rustproofing specialist have a good reputation? (Call the Better Business Bureau to find out.)

9. Is there a ceiling on the cost of the repair? A deductible amount?

10. If you move to a different city or state, can you get warrantied service there?

11. Is the rustproofer liable even when the manufacturer warranties the vehicle against rust?

12. Are periodic inspections required? Is there any fee for cleaning before inspection? If you move to a different location, can you get your car inspected there?

13. Is the claim process spelled out clearly?

Finally, see the chart in Appendix 3, which lists some of the warranty provisions that five specific rustproofing specialists offer.

IF YOU HAVE TO FILE A CLAIM

From the moment you purchase after-market rustproofing protection, you should begin keeping written records of any transactions that take place. For example, when you take your car in for inspection, note the date, the name of the inspector and any comments that are made. Keep records of money you spend on routine maintenance, too. This includes the cost of replacing weatherstripping or other minor components. You may need these later, should there ever be a question concerning the quality of care your car has received.

If rust develops, follow the procedures outlined in the warranty for filing a claim. But file a claim in writing. For example, instead of notifying the warrantor by phone when you first notice the rust, write a letter and send it via certified mail (and keep a copy). Or notify the warrantor by phone *and* by mail.

If you are involved in a collision, again, notify the warrantor immediately and follow the procedures outlined in the warranty. Keep estimates and receipts for repair of the collision damage. That way you can prove you have taken care of the automobile and met all warranty requirements.

If you get no response from the warrantor, or if you get a negative response, then write a letter (and send it certified mail) to someone at the next highest level — corporate headquarters, for example. If you are still unhappy with the results, contact your Better Business Bureau or state attorney general. In many cases, these agencies will have already received complaints from other consumers with the same problem. But even if they haven't, they can give you some basic advice, which may include a recommendation that you sue in small claims court.

If your warranty fails to outline the procedures for filing a claim, the best thing to do is to notify the warrantor in writing (certified mail again) whenever there is the slightest problem. That way you may be covered later, if you are told you did not meet a particular requirement.

Chapter 10
Factory Warranties

Warranties against corrosion are more extensive than ever. For its 1989 models, Chrysler offers seven-year coverage and Ford and General Motors are up to six years. Porsche offers its car owners 10-year coverage, regardless of mileage. Within the next couple of years, American carmakers may warrantee cars for 10 years as well. Does this mean rust will no longer be a car owner's worry?

I'm afraid not.

The terms of a warranty are decided not by a model's engineers, but by the marketing specialists at the different manufacturers. A warranty is not a statement of a particular model's corrosion resistance. It is merely a marketing device. And big warranties sell lots of cars.

But how can the manufacturers afford such warranties if they don't reflect the car's durability?

In several ways. First, they assume (and rightly so) that most car buyers will not take the time to read the warranty thoroughly, but will simply accept the dealer's explanation of its terms; many will not inquire about the warranty at all. In addition, many dealers neglect to even mention the factory warranty because they are trying to sell additional rustproofing.

Second, few prospective buyers sit down and go over the specific provisions of the different warranties, so it isn't until after the car has been purchased that they learn about the different types of corrosion that occur, and which of them are covered. For instance, did you know that usually only rust attacking a car from the inside out is covered, and that the owner is responsible for maintaining the paint finish?

Third, the manufacturers assume (and experience has borne them out) that a good number of new-car owners will fail to meet the terms of the warranty, which may include using only factory-approved replacement parts (which themselves must be rustproofed within a certain period) and washing the underbody regularly.

Fourth, research has shown that many people buying new cars will sell them within three or four years, and that the people buying these used cars will fail to take the steps necessary to transfer the corrosion warranty to their names. Moreover, some carmakers charge a fee (as much as $100) to make such a transfer.

Finally, the carmakers hide the cost of servicing the warranties in the cost of the cars themselves.

Nevertheless, a new-car warranty is a promise. It's a contract between the manufacturer and the consumer, a written guarantee that the car will perform as it was designed to for a specified amount of time. Like most legal agreements, it includes restrictions and specific requirements.

It is also something you should understand both before and after you buy a new car. You should understand warranties before you buy so you can compare the offerings of different manufacturers. You should understand them after you buy so you can get service and parts free of charge

WHAT A WARRANTY PROVIDES

Each new-car warranty offers three types of coverage: basic, power train, and rust perforation. Basic coverage ensures that the car (except for items like tires, oil filters, fan belts, and the like) will be free from defects, usually for a period of about 12 months or 12,000 miles.

Power train coverage ensures that the engine and transmission and other primary components will operate as designed, usually for a period of six or seven years (for the American automakers).

And rust perforation coverage guarantees that inside-out corrosion will be repaired, again for a period of six to seven years. This chapter concerns only rust perforation coverage.

Reading the fine print. Before you buy a car, ask to read the factory warranty. Focus especially on the section that lists warranty exclusions, or parts of the car that are *not* covered. This section will usually explain situations or circumstances that can invalidate the warranty. For example, among the exclusions listed on the warranty for a 1987 Japanese model are the following:

> Any perforation due to corrosion ... caused by industrial fallout, accident, damage, abuse or vehicle modification. Any surface corrosion ... which does not result in perforation, such as that typically caused by sand, salt, hail or stones. Any perforation due to corrosion ... which results not from a defect in material or workmanship, but from failure to maintain the vehicle in accordance with the procedures specified in the Owner's Manual provided ....

Almost all warranties exclude rust that develops from paint chips, scratches, dents or dings. Most also require the owner to maintain the car's finish by repairing all cosmetic damage and regularly washing the car's underbody. In addition, any area of the car damaged in a collision must be repaired with factory-approved materials and methods or warranty protection may be

1989 RUST WARRANTIES

Acura . . . . .3 years, unlimited mileage

Audi . . . . .6 years, unlimited mileage

Audi Advantage . . . . .10 years, unlimited mileage

BMW . . . . .6 years, unlimited mileage

Chrysler . . . . .7 years or 100,000 miles

Ford . . . . .6 years or 100,000 miles

General Motors . . . . .6 years or 100,000 miles

Honda . . . . .3 years or 36,000 miles

Hyundai . . . . .3 years, unlimited mileage

Isuzu . . . . .3 years, unlimited mileage

Jaguar . . . . .6 years, unlimited mileage

Jeep/Eagle . . . . . . .7 years or 100,000 miles

Mazda . . . . .5 years, unlimited mileage

Mercedes-Benz . . . . .4 years or 50,000 miles

Mitsubishi . . . . . . .5 years or 50,000 miles

Nissan . . . . .5 years, unlimited mileage

Porsche . . . . .10 years, unlimited mileage

Saab . . . . .6 years, unlimited mileage

Subaru . . . . .5 years or 60,000 miles

Toyota . . . . .5 years, unlimited mileage

Volkswagen . . . . . .6 years, unlimited mileage

Volvo . . . . .8 years, unlimited mileage

canceled. Moreover, these damaged areas may be permanently excluded from coverage, even after they have been repaired and approved by the dealer.

Given all these restrictions, it's easy to envision disputes over the cause of the corrosion needing repair. Was it owner neglect? If so, you pay for any repairs. Or was it faulty materials or workmanship? If it was, the manufacturer pays.

If warrantied service is denied. Before you find yourself in this position, make it a point to keep careful records of all maintenance and repairs involving your car. This means keeping all receipts and, perhaps, even a detailed log book of regular maintenance checks. It also means writing down the names of the people at the dealership who work on your car, along with any comments they make about the car's condition.

In addition, before you take your car in for a service appointment, read the warranty again so it will be fresh in your mind. If the dealer refuses to perform the service free of charge, and if your arguments fail to persuade him, take the matter to the manufacturer's regional or zone office.

These steps are probably outlined in your owner's manual, including the address and phone number of the manufacturer's office for your region. It is probably best to *write* to the regional office — instead of contacting the office by telephone — and to send your letter certified mail. A representative will then contact both you and the dealer and attempt to resolve the problem.

If you still do not get satisfaction, write to the manufacturer's customer relations department at headquarters (this address will undoubtedly be listed in the owner's manual). Write as objectively as possible, listing events as they occurred (chronologically). Don't accuse people of intending to do you wrong. Simply point out the services to which you are entitled, and explain when and by whom they were denied. Don't use emotional language; just write calmly. But do include the names of people within the company with whom you have spoken. Also include a copy of your letter to the regional office, and all documentation of maintenance and repairs. Send all this information certified mail.

If you still aren't pleased with the outcome, arbitration is the next step. This means an independent organization (such as Auto Line or Autocap) will intervene and rule on the dispute. If the organization rules in your favor, the ruling is binding. However, if it rules in the manufacturer's favor, you can still sue in court (usually small claims court). For more information on arbitration organizations, contact your local Better Business Bureau or state attorney general.

All new-car warranties are now limited warranties, not full warranties. This means you have no unstated, or implied, rights. Coverage is as it is written in the warranty, and it expires on the date listed in the warranty. However, some dealers will provide service free of charge even after the warranty has expired, depending on the problem at hand.

Secret warranty extensions. Several automakers — both American and foreign — offer "secret" warranty coverage. That means they sometimes provide service free of charge even after the factory warranty has expired. However, this free service is forthcoming only when the owners of particular models complain persistently. In other words, this free service is not publicized. Nor is it divulged upon direct questioning unless it is clear the car owner already knows about it.

For example, if a particular model has a problem with hesitation after starting, and if many car owners begin to complain about the problem, the manufacturer will sometimes direct its customer service personnel to fix the problem free of charge. Since these directives are issued through customer service — and not the dealer — the manufacturers bypass a federal rule requiring them to notify the National Highway Traffic Safety

Administration of any defects being repaired free of charge. As a result, no public notification program is initiated, and the manufacturers' reputations remain unblemished.

For this reason, the most important things you can do to protect yourself are to keep records and complain. Follow the steps listed above for writing to the manufacturer's regional offices and headquarters. You can also contact the Center for Auto Safety (2001 S St., N.W., Suite 410, Washington, D.C. 20009) if you want to find out whether any secret warranties exist for your car.

The chart on page 168 shows what the different carmakers offer in the way of corrosion warranties for their 1989 models.

Chapter 11
Rust & Auto Safety

Cosmetic corrosion usually gets immediate attention. After all, scratches, chipped paint, and dents and dings are ugly. They also reduce a car's resale value. But the corrosion that really matters — when it comes to safety — is slow-breeding and hidden. And it does a lot of damage before it ever becomes apparent to the average car owner. The most dangerous kind of corrosion is what I call "structural corrosion" — rust or decay that eats away vital supports.

Once it attacks the structure of your car, rust will aggravate the severity of any collision you have. In fact, it can cause as much damage as a faulty steering or suspension system can. It usually begins as a small, almost imperceptible, rust spot in a closed box section, where structural integrity is crucial. Then it spreads to other structural components. In the end, it leaves the car unable to sustain sudden shocks, like those generated in a collision.

Older cars are most susceptible to this kind of corrosion. But even newer cars are vulnerable just beneath the floors, along the rocker panels, and at the junctions between the floor and the firewall and door pillars.

System failure. In collisions involving older cars, as you may know, brake failure is often a contributing factor. But did you know that corrosion often contributes to brake failure? Brake systems corrode internally because brake fluid absorbs moisture over time. And brake *lines* corrode because they are exposed to water, mud, salt and slush kicked up by the tires.

Check these components regularly. Also drain the brake fluid periodically, replacing it with silicone fluid or regular brake fluid from a new, unopened can.

Front-end collisions. Collisions involving cars structurally weakened by rust follow specific patterns. In front-end collisions, the first significant masses to get forced back toward the passenger compartment are the engine and transmission. Sometimes the wheels and tires may get shoved back as well. Since these components are so massive, they will overload a corroded structure as they press against the firewall and front fenders. In serious collisions, these masses may move into the passenger "survival space" — the part of the passenger compartment designed to remain free and clear of obstruction — decreasing it by as much as 15 percent.

Another vulnerable area is the seat-to-floor connection, particularly in older cars. If it has been weakened by corrosion, it can fail during emergency braking. As a result, the seat may tear loose, and the passenger may get thrown around. The safety-belt-to-floor connection can fail under the same circumstances.

Doors may open under impact if the pillars are corroded, and the passenger may be thrown into oncoming traffic.

Side impacts. In side impacts, weakened rocker and floor panels can be hazardous. The colliding car can cut through these areas when they are badly rusted, severing the front and rear ends. In fact, in especially violent impacts, the colliding car can penetrate the passenger survival space by as much as 14 inches.

So don't take safety too lightly when you buy a car, particularly an older car. Have an expert check its structural components thoroughly. If you have had the same car for several years, be sure to have it examined periodically for structural corrosion. Your life may be at stake!

GOVERNMENT INTERVENTION

The U.S. Department of Transportation established the National Highway Traffic Safety Administration (NHTSA) to investigate complaints about manufacturing defects that may jeopardize auto safety. These defects can involve electronic, mechanical, or structural parts. The NHTSA operates a toll-free hotline so consumers can report safety problems or request recall information on any automobile make and model. After it investigates a safety problem, the NHTSA may begin a recall campaign or help the consumer resolve the problem with the manufacturer.

One NHTSA spokesperson said that, because of the public's demand for more corrosion-resistant cars, most automakers have extended the factory warranty against perforation and increased the rust-resistance of steel sheet metal in the factory. As a result, most of the safety complaints the NHTSA receives concerning new cars are *not* related to corrosion.

The spokesperson also said that the automakers are much more cooperative these days. That is, when the NHTSA points to a specific problem a consumer has encountered, the manufacturer usually takes the lead in proposing a solution. Nevertheless, there have been recalls — voluntary and involuntary, on the part of the automakers — related to corrosion.

About 10 years ago, 936,744 Honda Civics, CVCCs and station wagons, Accords, and Preludes were recalled when the government determined that exposure to road salts could cause structural failure of the undercarriage. The manufacturer was required to treat the undercarriage with rust preventives.

In 1980 and 1981, Fiat voluntarily recalled several of its models because the rear axle was falling out from under them — a corrosion problem.

Renault's 1983-85 Alliance and Encore models were recalled when it was determined that road salts could damage the steering-gear system. It was feared that drivers might lose control of their vehicles.

In addition, some 1971 and 1972 Chevrolet Vegas had fenders that quickly rusted through. While there was no recall campaign *per se,* the manufacturer did make some "policy adjustments" on a case-by-case basis (meaning it repaired some fenders free of charge).

All this should go to show you the value of complaining. If an area of your car seems especially prone to rust, complain about it. Complain to your dealer. Then complain to the manufacturer's regional office or the NHTSA.

Secret recalls. Sometimes the manufacturers have "secret" repair policies. That means they sometimes make repairs free of charge for problems with a particular model, without the NHTSA's involvement. For more information on these secret recalls, see Chapter 10.

By the way, the number for the NHTSA hotline is 800-424-9393. (In Washington, D.C., the number is 426-0123.)

Appendix 1
Useful Products

RUSTPROOFING SUPPLIES

"Professional" rustproofing kit

J.C. Whitney
See "Catalogues & Publications" below

$35.96. Part No. 97-5403NF.

Includes a 20-inch rigid spray wand, a 30-inch "superflex" wand and a 6-inch underbody wand, along with three different spray tips and illustrated instructions. You provide the air compressor and chemicals.

Rustproofing chemicals

Wurth USA
93 Grant St.
Ramsey, NJ 07446
(800) 526-5228
(201) 825-2710 in New Jersey

1 liter — $12.50. No. 892080.

A wax-based rust preventive for inner body cavities that will "destroy condensation films, penetrate rust and prevent further corrosion."

BODY TOOLS & KITS

10-piece auto body tool and repair kit

J.C. Whitney
See "Catalogues & Publications" below

$22.49. Part No. 88-8158T.

Includes a rubber sanding block, sandpaper, three body filler applicators, a finishing hammer, a dolly, a dent puller, a file-blade holder and two blades.

Panel flangers

Eastwood Co.
See "Catalogues & Publications" below

$29.95. Part No. 6286.

This tool, mentioned in Chapter 4, produces an offset flange so replacement panels fit flush. Adjustable to different metal gauges.

Hide-a-Screw tool

Technical Products Marketing
316 Cayton Bldg.
812 Huron Road
Cleveland, OH 44115
(216) 696-7877

$9.70.

An inexpensive version of dimpling pliers (see Chapter 4).

Mr. Fix de-rusting tool

Marson Corp.
130 Crescent Ave.
Chelsea, MA 02150
(617) 884-7760

Approximately $4.99. No. 81330.

This "pen" contains a bundle (approximately 1/8 inch thick) of fiberglass fibers that will remove rust from a small paint chip without damaging the good paint around it. It is also good for cleaning electrical contacts.

Flexible claw & magnetic pick-up

J.C. Whitney
See "Catalogues & Publications" below

Two 24-inch tools — $1.61. Part No. 81-2605W.

These tools are intended to pick up nuts and bolts that fall into hard-to-reach places, but they will also retrieve bits of rusty metal from interior sections (see p. 117).

Badger basic spray gun with propel can

Badger Air-Brush Co.
9128 W. Belmont Ave.
Franklin Park, IL 60131
(312) 678-3104

$20 suggested retail price. Model No. 250-3. Comes with an 11 oz. can of propellant.

Excellent for touching-up chips and other defects because it allows you to work close to the surface and repair very small areas without drips. This company sells a whole line of air brushes.

Body Man Tool

Eastwood Co.
See "Catalogues & Publications" below

$29.95. Part No. 6799.

This is the "dent removal tool" mentioned in Chapter 4. With it, you can pull out dents using a 3/8-inch power drill in the same time it would take to drill a hole for a slide hammer.

Panel holding system

Eastwood Co.
See "Catalogues & Publications" below

13-piece system — $29.95.

Reusable holders keep panels in place during welding and other repairs.

EPOXIES & ADHESIVES

POR-15 Putty

Stan Coleman Inc.
PO Box 1235
Morristown, NJ 07960
(800) 526-0796
(201) 887-1999 in New Jersey

1 pound — $10.

This two-part "construction material" has a clay-like consistency. It can be sanded, drilled or machined. The manufacturer supplies specific directions for use on aluminum, copper, brass, ferrous alloys, nylon, ABS, polycarbonate, stainless steel, chrome, stone and ceramics.

2-ton epoxy

Devcon Corp.
30 Endicott St.
Danvers, MA 01923
(508) 777-1100

1 oz. tube — $2.64. 1 oz. plastic syringe — $3.11. 9 oz. bottle — $11.

This product is waterproof, not just water-resistant. In general, the longer the curing time for an epoxy, the stronger its dry strength.

WELDING

Panel welder

Eastwood Co.
See "Catalogues & Publications" below

Welder — $199. No. 4373. Set of 1/16 inch arc welding rods — $19.95. No. 4374.

This low-amperage welder was described in Chapter 4. Comes with arc welding attachment, ground cord, power cord, face shield and instructions.

Wire Feed Welder

Eastwood Co.
See "Catalogues & Publications" below

$349.95. No. 7127.

Welds body metal or heavier steel up to 3/16 inch thick. Also available for this machine is an MIG gas conversion kit ($89.95). The Eastwood Co. also sells a wide range of welding products, including protective gloves, instructive books, welding attachments, and electrodes.

BODY PANELS

Tabco replacement panels and parts

Tabco
30500 Solon Industrial Pkwy.
Cleveland, OH 44139
(216) 248-5151

Call for free catalogue. This company sells a host of steel parts and panels for domestic and imported cars.

GENERAL CHEMICALS

Oxi-Solv degreaser

Solv-O Corp.
6995 Monroe Blvd.
Taylor, MI 48180
(313) 292-2060

Cleans grease and oil from iron, steel, aluminum, brass, bronze and chrome. This product is nontoxic, nonflammable and biodegradable. It does contain acid, however, so hand and eye protection should be worn.

Oxi-Solv rust remover

Eastwood Co.
See "Catalogues & Publications" below

16 oz. — $8.95 (No. 3430). 1 gallon — $24.95 (No. 3432).

See Chapter 6 for more details. This was my choice of the rust removers. The makers of Oxi-Solv (Solv-O Corp.) will probably only ship large orders, but Eastwood sells the volumes listed above.

Penetrant

3M Corp.
Automotive Trades Division
St. Paul, MN 55144
(612) 733-3300

Penetrates and loosens rusted nuts, bolts, screws and other parts. Also helps prevent corrosion.

Silicone brake fluid

Eastwood Co.
See "Catalogues & Publications" below

1 quart — $12.95.

Unlike regular brake fluids, silicone fluids do not absorb water. Consequently, system corrosion is substantially reduced.

Touch-up paint

Dupli Color Products Co.
1601 Nicholas Blvd.
Elk Grove Village, IL 60007
(312) 439-0600

Spray — $4.58 (No. 502). 1/2 oz. scratch fix with brush — $3.46.

Touch-up paints for import and domestic models from 1972 to present.

INSPECTION AIDS

Spot Rot[TM] *rust detector*

Pro Motorcar Products Inc.
1437 Gulf to Bay
Clearwater, FL 34615
(813) 447-0287

Consumer version — $12.95. Pro gauge — $29.95.

This product can also detect excessive layers of paint. In fact, the pro version can measure paint thicknesses to 0.0005 inch. Also useful for detecting clipped cars and showroom repaint jobs. Comes with an illustrated booklet explaining where to look for rust and how to buy a used car.

Home-made body filler detectors

You can detect rust (or collision damage) that has been treated with body filler with a device made in the following manner. Place several layers of shirt cardboard between a small magnet and a vertical surface of your car. Add cardboard until the magnet would no longer stick if you added just one more piece. Then tape the cardboard and magnet together.

MISCELLANEOUS

Rust Evader electronic cathodic device

J.C. Whitney
See "Catalogues and Publications" below

$146.69. Part No. 12-6201U.

This product, described in Chapter 3, is available through some new-car dealerships and J.C. Whitney. Prices vary widely among dealerships.

"Stop Radiator Rust!" device

J.C. Whitney
See "Catalogues & Publications" below

$4.46. Part No. 74-7496U.

A "sacrificial" zinc anode for the radiator. This device fits beneath the radiator cap, where it prevents corrosion in the cooling system.

TV/Radio cable or connector sealant

Radio Shack/Tandy Corp.
PO Box 17180
Fort Worth, TX 76102
(817) 390-3700

$2.49. Part No. 278-1645.

This is the sealant mentioned in Chapter 3 for waterproofing electrical connections.

Universal door, trunk and hatch weatherstripping

J.C. Whitney
See "Catalogues & Publications" below

13-1/2 foot length — $13.49. Part No. 81-9909N.

J.C. Whitney also sells weatherstripping for specific makes and models. See its catalogue.

Oxi-Solv rust spot remover kit

Solv-O Corp.
6995 Monroe Blvd.
Taylor, MI 48180
(313) 292-2060

8 oz. — $9.95. 16 oz. — $13.95.

This kit includes plastic film pockets, which contain absorbent pads backed with an adhesive film. The pockets are stuck to the rusted area, filled with Oxi-Solv (see Chapter 6) and then sealed. The pockets keep vertical surfaces wet during rust treatment. You can make them yourself with plastic film, tape and a rag.

Turbo-Wash

Turbo Tek Inc.
PO Box 34081
888 Venice Blvd., Suite 205
Los Angeles, CA 90034
(213) 559-5540

Approximately $15.

A spray washer that connects to a garden hose. Comes with three spray tips and a soap/water ratio control knob. Two of the tips are suitable for the underbody and wheel wells.

Carnauba wax

Malm Chemical Corp.
Box 300 CCMM
Pound Ridge, NY 10576
(914) 764-5775

1 pint — $16.95.

100 percent nonabrasive wax safe for use on the newest finishes. This company also sells polishes and compounds.

Auto Price Service

Consumers Union/Consumer Reports
PO Box 570
Lathrup Village, MI 48076

Approximately $10 per model.

The printouts from this service list all standard equipment and all options available for new cars by make and model, along with invoice prices (what the dealer pays the manufacturer). They are a boon to anyone shopping for a new car at the best price.

CATALOGUES & PUBLICATIONS

Eastwood Co.

580 Lancaster Ave.
PO Box 296
Malvern, PA 19355
(215) 640-1450
Order line: (800) 345-1178

Eastwood bills itself as specializing in auto restoration tools and techniques. Its catalogue offers many original items for restoration and general maintenance. In addition, Eastwood's customer service (the 800 phone number) answers questions about products it sells. The company also offers a full refund for 30 days after purchase.

J.C. Whitney & Co.

1917-19 Archer Ave.
PO Box 8410
Chicago, IL 60680
Order line: (312) 431-6102

This catalogue lists thousands of items, from automotive supplies in general to special sections for VW, Karmann Ghia, Jeep, Bronco, Toyota, IHC Scout, Camaro, Firebird, Mustang, Corvette, vans and campers, antique and imported cars, and early-model Fords and Chevrolets. Most of the items are discounted, but J.C. Whitney usually does not reveal the names of the manufacturers of the different products it sells.

U.S. General

100 Commercial St.
Plainview, NY 11803
(516) 349-7275
(800) 645-7077

This catalogue includes a variety of tools including spray guns, air brushes, compressors, sandblasters and high-pressure washers.

Motobooks

Albion Scott
48 E. 50th St.
New York, NY 10022-6821
(212) 980-1928
Or 3800 East Coat Highway
Corona Del Mar, CA 92625
(714) 673-7007

General-interest and specialized automotive books by mail.

Cars & Parts

PO Box 482
Sidney, OH 45365
(513) 498-0803

This monthly magazine includes articles on restoration regularly. It also includes ads for restoration supplies and parts, and cars and parts.

Car Collector & Car Classics

8601 Dunwoody Place, Suite 144
Atlanta, GA 30350
(404) 998-4603

A monthly magazine published by Classic Publishing Inc.

Collectible Automobile

Publications International Ltd.
3841 W. Oakton
Skokie, IL 60076

A bimonthly magazine featuring illustrated articles about past and present classics.

Hemmings' Motor News

Box 100
Bennington, VT 05201

In Canada, contact Frontier Distributing, PO Box 1051, Fort Erie, Ontario, L2A 5N8. Hemmings calls itself "the world's largest antique, vintage and special interest auto marketplace." It lists events and auctions, books, tools, new and used car parts, and cars themselves from 1903 on. A sister publication is *Hemmings' Vintage Auto Almanac*, which lists parts suppliers, restoration shops, old-car dealers, salvage yards and car clubs.

Old Cars Weekly

700 E. State St.
Iola, WI 54990
(715) 445-2214

Aside from the magazine, this company also publishes the *Old Car Price Guide 1901-1979.*

Practical Classics & Car Restorer

For U.S. subscriptions contact:
Eric Walter Associates
Gleneagles Dr.
New Vernon, NJ 07976
(201) 267-5612

A monthly magazine published in England that focuses on restoration and restoration products. Interesting even if you don't own a British car.

Skinned Knuckles

175 May Ave.
Monrovia, CA 91016
(818) 358-6255

A monthly journal of car restoration.

Special Interest Autos

Box 196
Bennington, VT 05201

A bimonthly collector-car magazine that features how-to articles on restoration. It also contains frequent features on collectables and related information, such as specification charts.

Appendix 2

Chemicals & Coatings
Product Information

RUST CONVERTERS

Product	Package	Shelf life	Toxic/flammable	Recoat/topcoat
EXTEND Loctite Corp. 4450 Cranwood Ct Cleveland, OH 44128 (216) 475-3600	Aerosol	2+ years	Skin/eye irritant Extremely flammable	30-60 minutes recoat Overnight topcoat
RUST REFORMER Rust-Oleum 11 Hawthorn Pkwy Vernon Hills, IL 60061 (312) 367-7700	Liquid	5+ years	Skin/eye irritant Nonflammable	3 hours recoat 3 hours topcoat
NEUTRA RUST New York Bronze Powder Co. Elizabeth, NJ 07201 (201) 578-2000	Liquid	1 year	Skin irritant Nonflammable	2 hours recoat 72 hours topcoat

RUST CONVERTERS (Continued)

Product	Package	Shelf life	Toxic/flammable	Recoat/topcoat
TRUSTAN Trustan Co. 2094 Boston Post Rd Larchmont, NY 10538 (914) 834-0974	Liquid	1+ years	Skin/eye irritant Nonflammable	15 minutes to 1 hour recoat 12 to 24 hours topcoat (primer mandatory)
RUST AVENGER 3M Corp. Automotive Trades Div. 3M Center St. Paul, MN 55144-1000 (612) 733-1110	Liquid (as part of kit) or pen-shaped applicator (for nicks & chips)	3+ years	Skin irritant Nonflammable	30 minutes recoat 30 minutes topcoat

RUST CONVERTERS (Continued)

Product	Topcoats	Dry temp. range	Advantages	Disadvantages
EXTEND	No water-based or auto finishes	-35° to 200°F	—	Not for internal areas (e.g. rocker panels) or moisture pockets
RUST REFORMER	Epoxy, vinyl or polyurethane	No lower limit 200°F	—	—
NEUTRA RUST	Standard or syn-thetic lacquers; oil-based paints; two-component lacquers or bitumastic & oil tar-based finishes	-4° to 300°F	—	Begins to degrade at 250°F

RUST CONVERTERS (Continued)

Product	Topcoats	Dry temp. range	Advantages	Disadvantages
TRUSTAN	New acrylics; epoxy; bituminous coatings; oil; enamel & lacquer	No limit if topcoated	—	Will flash-rust if not topcoated
RUST AVENGER	Enamel & lacquer if primed	Not available, but product has weathered 3 Minnesota winters	Sandable wet or dry Available in 2 applicators: pen & bottle	—

RUST REMOVERS

Product	Shelf life	No. of applications	Weldable?	Other
OXI-SOLV Solv-O Corp. PO Box 787 Taylor, MI 48180 (313) 292-2060	Indefinite	Apply & rinse; repeat, if necessary	Must be rinsed from surface first	Author's choice among rust removers

RUST REMOVERS (Continued)

Product	Shelf life	No. of applications	Weldable?	Other
CORTEC 420 & 421 Cortec Corp. 310 Chester St. St. Paul, MN 55107 (612) 224-5643	24 months	Apply & rinse; repeat, if necessary	Must be rinsed from surface first	Products contain phosphoric acid
CORTEC 422 & 423 Cortec Corp. (see above)	Not available because of products' newness	Apply & rinse; repeat, if necessary	Must be rinsed from surface first	Recently reintroduced; limited availability
RUST-A-HOY Travaco Labs Inc. 345 Eastern Ave. Chelsea, MA 02150 (617) 884-7740	Indefinite	Apply & rinse; repeat, if necessary	Must be rinsed from surface first	Contains phosphoric acid

RUST INHIBITORS

Product	Shelf life	No. of coats	Weldable?	Other
CORTEC VCI-319 Cortec Corp. 310 Chester St. St. Paul, MN 55107	24 months (protect from freezing temperatures)	1 thick coat	Yes; however, it emits black smoke, which is nontoxic but annoying	Wear a respirator during application
CORTEC VCI-376 Cortec Corp. (see above)	24 months (protect from freezing temperatures)	3-5 mil coating required	Yes; however it emits black smoke, which is nontoxic but annoying	With a topcoat, it is especially durable in harsh environments; hazardous for home use
COLD GALVANIZING COMPOUND Eastwood Co. PO Box 296 Malvern, PA 19355 (800) 345-1178	Aerosol: 1 year Liquid: 1+ years	1 thick coat	Yes	Requires sealer before topcoating Flammable Available to public

PRIMERS & PAINTS

Product	Shelf life	Health hazards	Compatible topcoats
7769 RUSTY METAL PRIMER Rust-Oleum 11 Hawthorn Parkway Vernon Hills, IL 60061 (312) 367-7700	5 years	Wear face mask during application (respirator in poorly ventilated areas)	Alkyds No auto finishes
7773 CLEAN METAL PRIMER Rust-Oleum (see above)	5+ years	Skin irritant; wear goggles, gloves & NIOSH respirator	Rust-Oleum's 7700 series paints
FSC-1451 RED PENETRATING PRIMER Lasting Paints Inc. 200 S. Franklintown Rd Baltimore, MD 21223 (301) 947-6300	Not available	Contains zinc chromate; overexposure can lead to serious respiratory problems; wear gloves, goggles & an air-line respirator	Industrial primers & paints; no auto finishes

PRIMERS & PAINTS (Continued)

Product	Shelf life	Health hazards	Compatible topcoats
POR-15 Stan Coleman Inc. PO Box 1235 Morristown, NJ 07960 (800) 526-0796	6 months[1] (at 77°F once can is opened)	Severe eye & respiratory irritation[2]; wear goggles, gloves & respirator	Compatible with regular automotive finishes
ALUM. EPOXY MASTIC 2 (FSC 901 & 902) Lasting Paints Inc. (see above)	Not available	Moderate to severe eye & respiratory irritation; wear goggles, gloves & respirator	Industrial primers & paints; no auto finishes

[1] *An unopened sample of the product stored indoors for one year began to bulge; only metal tabs kept the lid from coming off.*

[2] *The Threshold Limit Value (TLV) for isocyanate (an ingredient of POR-15) is 0.02 part per million. Because of POR-15's low volatility, high exposures are not anticipated unless the product is overheated or sprayed.*

Appendix 3

5 Rustproofing Warrantors
& Their Terms

5 RUSTPROOFING WARRANTORS & THEIR TERMS

Warrantor	Type of warranty	Duration	Transferable?	Applic. deadline
FORD SUPER SEAL	Limited	As long as you own your car	No	On or before delivery
POLY-GLYCOAT	Limited	7 years	Yes[1]	3,000 miles or 4 months
QUAKER STATE	Limited	As long as you own your car	No	3,000 miles or 90 days
RUSTY JONES	Full	As long as you own your car	Yes[2]	6 months
ZIEBART	Limited	As long as you own your car	Yes[3]	New car

Source: New York State Attorney General's Report, A Tarnished Option.

Note: Rusty Jones filed for bankruptcy in 1988, and some of the other warrantors listed above may have changed the terms of their coverage. Nevertheless, this table should illustrate the variety of terms available and the importance of reading the fine print closely.

5 WARRANTORS (Continued)

Warrantor	Inspections	Coverage	Exclusions	Remedies
FORD SUPER SEAL	Every 18 months; dealer may charge for cleaning	Hood; fenders; door & quarter panels; wheel housing; deck lid; lift & station wagon tail gates; floor pan; rocker panels; cowl	Outside-in rust	Repair or replace damage (warrantor's option); limit of warrantor's liability (not to exceed book value); 1 repair or cash settlement per car
POLY-GLYCOAT	No inspection required	All areas properly coated	Uncoated areas; collision damage; defects covered by new-car warranty; factory defects; chipped paint	Repair or refund of application cost
QUAKER STATE	Within 30 days of 3rd, 6th & 9th anniversaries of application[4]	Underbody sheet metal & exterior sheet metal from bottom of window	Surface rust; car areas that undergo rust or collision repair; in cases	Repair or replace damage (warrantor's option); limit of warrantor's liability

5 WARRANTORS (Continued)

Warrantor	Inspections	Coverage	Exclusions	Remedies
QUAKER STATE (Continued)		line down	where liability falls to manufacturer	not to exceed book value; 1 repair or replacement per car
RUSTY JONES	Annually[5] (inspection validation must be mailed to warrantor)	Any area that rusts through	Muffler, exhaust & surface rust from chipping, collision or chemical damage	Repair or refund of application cost at consumer's option
ZIEBART	Inspection required once in first 5 years; free inspection & retreatment	Inside-out rust from treated areas	Exterior rust; metal trim; noncoated areas; exhaust system; moving parts	Repair damaged area

(1) But owner must return original warranty for reissue to new owner; in addition, a $15 fee is required. The reissued warranty is for the balance of the 7-year term. (2) However, within 2 weeks of the transfer, the Certificate of Transfer must be taken to a transfer agent and filled out, and the transferee's remedies are limited. (3) The warranty is transferable within 3 years, subject to inspection and fee payment (no amount specified). The transfer is good for 6 years from the date of application. (4) No charge for inspection and touch-up. (5) The annual inspection must take place within 30 days of anniversary of original application.

Index

A

Trunk repair 57

U

Underbody (or undercarriage)
Inspection 14-15, 42, 140
Rustproofing 16
Steam-cleaning 15, 20, 42, 90
Washing 14
Undercoating
Factory & after-market 136
Removal 70
versus rustproofing 14, 44
Unibodies
Repairing damage 124
versus body-on-frame design 123-124
Vulnerability to rust 11
Used cars
Inspection checklist (box) 141, 143
Rustproofing 155

W

Warranties
After-market rustproofing 157
After-market rustproofing loopholes 158
Arbitration 170
Exclusions (factory) 167
Factory 137, 165
Factory versus after-market 160
Federal laws governing 152
Filing a claim (after-market rustproofing) 163
If service is denied 169
1989 rust warranties (factory) 168
Owner obligations 166-167
Secret extensions 170
Transfer of ownership 142, 159
Types of coverage 166
Washing a car 17
Wax
Application 26
As protection 20
One-step 21
Summary (box) 24
Two-step 21
Types of 22

Ward Hill Press

**40 Willis Avenue
Staten Island, NY 10301
Phone: (718) 816-9449**

Please send me a copy of the following:

Rust: How to Keep It From Destroying Your Car $19.95
218 pages

Home Body Shop: Painting Your Car $9.95
30 pages

I understand that I may return either publication within 30 days for a full refund if not satisfied.

Name: _____

Address: _____

City, State, Zip: _____

New York residents please add appropriate sales tax.

Shipping: One publication: $1.50
      Both publications: $2

Special Option: First Class Mail at $3 per publication

Ward Hill Press

**40 Willis Avenue
Staten Island, NY 10301
Phone: (718) 816-9449**

Please send me a copy of the following:

Rust: How to Keep It From Destroying Your Car   *$19.95*
218 pages

Home Body Shop: Painting Your Car   *$9.95*
30 pages

I understand that I may return either publication within 30 days for a full refund if not satisfied.

Name: _____

Address: _____

City, State, Zip: _____

New York residents please add appropriate sales tax.

Shipping: One publication: $1.50
                    Both publications: $2

Special Option: First Class Mail at $3 per publication